THOMAS HENRY HUXLEY

SCIENCE AND CULTURE AND OTHER ESSAYS

Elibron Classics
www.elibron.com

Elibron Classics series.

ISBN 1-4021-6186-7 (paperback)
ISBN 1-4021-0467-7 (hardcover)

This Elibron Classics Replica Edition is an unabridged facsimile of the edition published in 1888 by Macmillan and Co., London.

SCIENCE AND CULTURE

AND

OTHER ESSAYS

SCIENCE AND CULTURE

AND

OTHER ESSAYS

BY

THOMAS HENRY HUXLEY, LL.D., F.R.S.

London
MACMILLAN AND CO.
AND NEW YORK
1888

First Edition printed 1881
Reprinted 1882, 1888.

PREFACE.

The Addresses, Lectures, and Essays gathered together in this volume have appeared at intervals during the past seven years, and I can give no better reason for republishing them in their present form, than the fact that three earlier collections of a similar nature have been received with favour, and, indeed, have not yet ceased to be in request.

I beg leave to offer my best thanks to the Editors and Publishers of the various publications in which these pieces have appeared, for their kindly accorded permission to reprint them.

London. *October* 1881.

CONTENTS.

I.

SCIENCE AND CULTURE.

SIX years ago, as some of my present hearers may remember, I had the privilege of addressing a large assemblage of the inhabitants of this city, who had gathered together to do honour to the memory of their famous townsman, Joseph Priestley;[1] and, if any satisfaction attaches to posthumous glory, we may hope that the manes of the burnt-out philosopher were then finally appeased.

No man, however, who is endowed with a fair share of common sense, and not more than a fair share of vanity, will identify either contemporary or posthumous fame with the highest good; and Priestley's life leaves no doubt that he, at any rate, set a much higher value upon the advancement of knowledge, and the promotion of that freedom of thought which is at once the cause and the consequence of intellectual progress.

Hence I am disposed to think that, if Priestley could be amongst us to-day, the occasion of our meeting would afford him even greater pleasure than the proceedings which celebrated the centenary of his

[1] See *Joseph Priestley*, p. 94, *infra*.

chief discovery. The kindly heart would be moved, the high sense of social duty would be satisfied, by the spectacle of well-earned wealth, neither squandered in tawdry luxury and vainglorious show, nor scattered with the careless charity which blesses neither him that gives nor him that takes, but expended in the execution of a well-considered plan for the aid of present and future generations of those who are willing to help themselves.

We shall all be of one mind thus far. But it is needful to share Priestley's keen interest in physical science; and to have learned, as he had learned, the value of scientific training in fields of inquiry apparently far remote from physical science; in order to appreciate, as he would have appreciated, the value of the noble gift which Sir Josiah Mason has bestowed upon the inhabitants of the Midland district.

For us children of the nineteenth century, however, the establishment of a college under the conditions of Sir Josiah Mason's Trust, has a significance apart from any which it could have possessed a hundred years ago. It appears to be an indication that we are reaching the crisis of the battle, or rather of the long series of battles, which have been fought over education in a campaign which began long before Priestley's time, and will probably not be finished just yet.

In the last century, the combatants were the champions of ancient literature, on the one side, and those of modern literature on the other; but, some thirty years[1] ago, the contest became complicated by

[1] The advocacy of the introduction of physical science into general

the appearance of a third army, ranged round the banner of Physical Science.

I am not aware that any one has authority to speak in the name of this new host. For it must be admitted to be somewhat of a guerilla force, composed largely of irregulars, each of whom fights pretty much for his own hand. But the impressions of a full private, who has seen a good deal of service in the ranks, respecting the present position of affairs and the conditions of a permanent peace, may not be devoid of interest; and I do not know that I could make a better use of the present opportunity than by laying them before you.

From the time that the first suggestion to introduce physical science into ordinary education was timidly whispered, until now, the advocates of scientific education have met with opposition of two kinds. On the one hand, they have been pooh-poohed by the men of business who pride themselves on being the representatives of practicality; while, on the other hand, they have been excommunicated by the classical scholars, in their capacity of Levites in charge of the ark of culture and monopolists of liberal education.

The practical men believed that the idol whom they worship—rule of thumb—has been the source of the past prosperity, and will suffice for the future welfare of the arts and manufactures. They were of

education by George Combe and others commenced a good deal earlier; but the movement had acquired hardly any practical force before the time to which I refer.

opinion that science is speculative rubbish; that theory and practice have nothing to do with one another; and that the scientific habit of mind is an impediment, rather than an aid, in the conduct of ordinary affairs.

I have used the past tense in speaking of the practical men—for although they were very formidable thirty years ago, I am not sure that the pure species has not been extirpated. In fact, so far as mere argument goes, they have been subjected to such a *feu d'enfer* that it is a miracle if any have escaped. But I have remarked that your typical practical man has an unexpected resemblance to one of Milton's angels. His spiritual wounds, such as are inflicted by logical weapons, may be as deep as a well and as wide as a church door, but beyond shedding a few drops of ichor, celestial or otherwise, he is no whit the worse. So, if any of these opponents be left, I will not waste time in vain repetition of the demonstrative evidence of the practical value of science; but knowing that a parable will sometimes penetrate where syllogisms fail to effect an entrance, I will offer a story for their consideration.

Once upon a time, a boy, with nothing to depend upon but his own vigorous nature, was thrown into the thick of the struggle for existence in the midst of a great manufacturing population. He seems to have had a hard fight, inasmuch as, by the time he was thirty years of age, his total disposable funds amounted to twenty pounds. Nevertheless, middle life found him giving proof of his comprehension of the practical problems he had been roughly called upon to solve, by a career of remarkable prosperity.

Finally, having reached old age with its well-earned surroundings of "honour, troops of friends," the hero of my story bethought himself of those who were making a like start in life, and how he could stretch out a helping hand to them.

After long and anxious reflection this successful practical man of business could devise nothing better than to provide them with the means of obtaining "sound, extensive, and practical scientific knowledge." And he devoted a large part of his wealth and five years of incessant work to this end.

I need not point the moral of a tale which, as the solid and spacious fabric of the Scientific College assures us, is no fable, nor can anything which I could say intensify the force of this practical answer to practical objections.

We may take it for granted then, that, in the opinion of those best qualified to judge, the diffusion of thorough scientific education is an absolutely essential condition of industrial progress; and that the College which has been opened to-day will confer an inestimable boon upon those whose livelihood is to be gained by the practice of the arts and manufactures of the district.

The only question worth discussion is, whether the conditions, under which the work of the College is to be carried out, are such as to give it the best possible chance of achieving permanent success.

Sir Josiah Mason, without doubt most wisely, has left very large freedom of action to the trustees, to whom he proposes ultimately to commit the administration of the College, so that they may be able to adjust its arrangements in accordance with the changing conditions of the future. But, with respect to three points, he has laid most explicit injunctions upon both administrators and teachers.

Party politics are forbidden to enter into the minds of either, so far as the work of the College is concerned; theology is as sternly banished from its precincts; and finally, it is especially declared that the College shall make no provision for "mere literary instruction and education."

It does not concern me at present to dwell upon the first two injunctions any longer than may be needful to express my full conviction of their wisdom. But the third prohibition brings us face to face with those other opponents of scientific education, who are by no means in the moribund condition of the practical man, but alive, alert, and formidable.

It is not impossible that we shall hear this express exclusion of "literary instruction and education" from a College which, nevertheless, professes to give a high and efficient education, sharply criticised. Certainly the time was that the Levites of culture would have sounded their trumpets against its walls as against an educational Jericho.

How often have we not been told that the study of physical science is incompetent to confer culture; that it touches none of the higher problems of life;

and, what is worse, that the continual devotion to scientific studies tends to generate a narrow and bigoted belief in the applicability of scientific methods to the search after truth of all kinds. How frequently one has reason to observe that no reply to a troublesome argument tells so well as calling its author a "mere scientific specialist." And, as I am afraid it is not permissible to speak of this form of opposition to scientific education in the past tense; may we not expect to be told that this, not only omission, but prohibition, of "mere literary instruction and education" is a patent example of scientific narrow-mindedness?

I am not acquainted with Sir Josiah Mason's reasons for the action which he has taken; but if, as I apprehend is the case, he refers to the ordinary classical course of our schools and universities by the name of "mere literary instruction and education," I venture to offer sundry reasons of my own in support of that action.

For I hold very strongly by two convictions—The first is, that neither the discipline nor the subject-matter of classical education is of such direct value to the student of physical science as to justify the expenditure of valuable time upon either; and the second is, that for the purpose of attaining real culture, an exclusively scientific education is at least as effectual as an exclusively literary education.

I need hardly point out to you that these opinions, especially the latter, are diametrically opposed to those of the great majority of educated Englishmen, influenced as they are by school and university traditions. In their belief, culture is obtainable only by a liberal education; and a liberal education is synonymous, not merely with education and instruction in literature, but in one particular form of literature, namely, that of Greek and Roman antiquity. They hold that the man who has learned Latin and Greek, however little, is educated; while he who is versed in other branches of knowledge, however deeply, is a more or less respectable specialist, not admissible into the cultured caste. The stamp of the educated man, the University degree, is not for him.

I am too well acquainted with the generous catholicity of spirit, the true sympathy with scientific thought, which pervades the writings of our chief apostle of culture to identify him with these opinions; and yet one may cull from one and another of those epistles to the Philistines, which so much delight all who do not answer to that name, sentences which lend them some support.

Mr. Arnold tells us that the meaning of culture is "to know the best that has been thought and said in the world." It is the criticism of life contained in literature. That criticism regards "Europe as being, for intellectual and spiritual purposes, one great confederation, bound to a joint action and working to a common result; and whose members have, for their common outfit, a knowledge of Greek, Roman, and Eastern antiquity, and of one another. Special, local, and temporary advantages being put out of account, that modern nation will in the intellectual and spiritual sphere make most progress, which most thoroughly carries out this programme. And what is that but saying that we too, all of us, as individuals, the more thoroughly we carry it out, shall make the more progress?"[1]

We have here to deal with two distinct propositions. The first, that a criticism of life is the essence of culture; the second, that literature contains the materials which suffice for the construction of such a criticism.

I think that we must all assent to the first proposition. For culture certainly means something quite different from learning or technical skill. It implies the possession of an ideal, and the habit of critically estimating the value of things by comparison with a theoretic standard. Perfect culture should supply a complete theory of life, based upon a clear knowledge alike of its possibilities and of its limitations.

But we may agree to all this, and yet strongly dissent from the assumption that literature alone is competent to supply this knowledge. After having learnt all that Greek, Roman, and Eastern antiquity have thought and said, and all that modern literatures have to tell us, it is not self-evident that we have laid a sufficiently broad and deep foundation for that criticism of life which constitutes culture.

Indeed, to any one acquainted with the scope of physical science, it is not at all evident. Considering progress only in the "intellectual and spiritual sphere,"

[1] *Essays in Criticism*, p. 37.

I find myself wholly unable to admit that either nations or individuals will really advance, if their common outfit draws nothing from the stores of physical science. I should say that an army, without weapons of precision and with no particular base of operations, might more hopefully enter upon a campaign on the Rhine, than a man, devoid of a knowledge of what physical science has done in the last century, upon a criticism of life.

When a biologist meets with an anomaly, he instinctively turns to the study of development to clear it up. The rationale of contradictory opinions may with equal confidence be sought in history.

It is, happily, no new thing that Englishmen should employ their wealth in building and endowing institutions for educational purposes. But, five or six hundred years ago, deeds of foundation expressed or implied conditions as nearly as possible contrary to those which have been thought expedient by Sir Josiah Mason. That is to say, physical science was practically ignored, while a certain literary training was enjoined as a means to the acquirement of knowledge which was essentially theological.

The reason of this singular contradiction between the actions of men alike animated by a strong and disinterested desire to promote the welfare of their fellows, is easily discovered.

At that time, in fact, if any one desired knowledge beyond such as could be obtained by his own observation, or by common conversation, his first necessity was to learn the Latin language, inasmuch as all the higher knowledge of the western world was contained in works written in that language. Hence, Latin grammar, with logic and rhetoric, studied through Latin, were the fundamentals of education. With respect to the substance of the knowledge imparted through this channel, the Jewish and Christian Scriptures, as interpreted and supplemented by the Romish Church, were held to contain a complete and infallibly true body of information.

Theological dicta were, to the thinkers of those days, that which the axioms and definitions of Euclid are to the geometers of these. The business of the philosophers of the middle ages was to deduce from the data furnished by the theologians, conclusions in accordance with ecclesiastical decrees. They were allowed the high privilege of showing, by logical process, how and why that which the Church said was true, must be true. And if their demonstrations fell short of or exceeded this limit, the Church was maternally ready to check their aberrations, if need be, by the help of the secular arm.

Between the two, our ancestors were furnished with a compact and complete criticism of life. They were told how the world began and how it would end; they learned that all material existence was but a base and insignificant blot upon the fair face of the spiritual world, and that nature was, to all intents and purposes, the playground of the devil; they learned that the earth is the centre of the visible universe, and that man is the cynosure of things terrestrial; and more especially was it inculcated that the course of nature had no fixed order, but that it could be, and constantly was, altered by the agency of innumerable spiritual beings, good and bad, according as they were moved by the deeds and prayers of men. The sum and substance of the whole doctrine was to produce the conviction that the only thing really worth knowing in this world was how to secure that place in a better which, under certain conditions, the Church promised.

Our ancestors had a living belief in this theory of life, and acted upon it in their dealings with education, as in all other matters. Culture meant saintliness—after the fashion of the saints of those days; the education that led to it was, of necessity, theological; and the way to theology lay through Latin.

That the study of nature—further than was requisite for the satisfaction of everyday wants—should have any bearing on human life was far from the thoughts of men thus trained. Indeed, as nature had been cursed for man's sake, it was an obvious conclusion that those who meddled with nature were likely to come into pretty close contact with Satan. And, if any born scientific investigator followed his instincts, he might safely reckon upon earning the reputation, and probably upon suffering the fate, of a sorcerer.

Had the western world been left to itself in Chinese isolation, there is no saying how long this state of things might have endured. But, happily, it was not left to itself. Even earlier than the thirteenth century, the development of Moorish civilisation in Spain and the great movement of the Crusades had introduced the leaven which, from that day to this, has never ceased to work. At first, through the intermediation of Arabic translations, afterwards, by the study of the originals, the western nations of Europe became acquainted with the writings of the ancient philosophers and poets, and, in time, with the whole of the vast literature of antiquity.

Whatever there was of high intellectual aspiration or dominant capacity in Italy, France, Germany, and England, spent itself for centuries in taking possession of the rich inheritance left by the dead civilisations of Greece and Rome. Marvellously aided by the invention of printing, classical learning spread and flourished. Those who possessed it prided themselves on having attained the highest culture then within the reach of mankind.

And justly. For, saving Dante on his solitary pinnacle, there was no figure in modern literature at the time of the Renascence to compare with the men of antiquity; there was no art to compete with their sculpture; there was no physical science but that which Greece had created. Above all, there was no other example of perfect intellectual freedom—of the unhesitating acceptance of reason as the sole guide to truth and the supreme arbiter of conduct.

The new learning necessarily soon exerted a profound influence upon education. The language of the monks and schoolmen seemed little better than gibberish to scholars fresh from Virgil and Cicero, and the study of Latin was placed upon a new foundation. Moreover, Latin itself ceased to afford the sole key to knowledge. The student who sought the highest thought of antiquity, found only a second-hand reflection of it in Roman literature, and turned his face to the full light of the Greeks. And after a battle, not altogether dissimilar to that which is at present being fought over the teaching of physical science, the study of Greek was recognised as an essential element of all higher education.

Thus the Humanists, as they were called, won the day; and the great reform which they effected was of incalculable service to mankind. But the Nemesis of all reformers is finality; and the reformers of education, like those of religion, fell into the profound, however common, error of mistaking the beginning for the end of the work of reformation.

The representatives of the Humanists, in the nineteenth century, take their stand upon classical education as the sole avenue to culture, as firmly as if we were still in the age of Renascence. Yet, surely, the present intellectual relations of the modern and the ancient worlds are profoundly different from those which obtained three centuries ago. Leaving aside the existence of a great and characteristically modern literature, of modern painting, and, especially, of modern music, there is one feature of the present state of the civilised world which separates it more widely from the Renascence, than the Renascence was separated from the middle ages.

This distinctive character of our own times lies in the vast and constantly increasing part which is played by natural knowledge. Not only is our daily life shaped by it, not only does the prosperity of millions of men depend upon it, but our whole theory of life has long been influenced, consciously or unconsciously, by the general conceptions of the universe, which have been forced upon us by physical science.

In fact, the most elementary acquaintance with the results of scientific investigation shows us that they offer a broad and striking contradiction to the opinions so implicitly credited and taught in the middle ages.

The notions of the beginning and the end of the world entertained by our forefathers are no longer credible. It is very certain that the earth is not the chief body in the material universe, and that the world is not subordinated to man's use. It is even more certain that nature is the expression of a definite order with which nothing interferes, and that the chief business of mankind is to learn that order and govern themselves accordingly. Moreover this scientific "criticism of life" presents itself to us with different credentials from any other. It appeals not to authority, nor to what anybody may have thought or said, but to nature. It admits that all our interpretations of natural fact are more or less imperfect and symbolic, and bids the learner seek for truth not among words but among things. It warns us that the assertion which outstrips evidence is not only a blunder but a crime.

The purely classical education advocated by the representatives of the Humanists in our day, gives no inkling of all this. A man may be a better scholar than Erasmus, and know no more of the chief causes of the present intellectual fermentation than Erasmus did. Scholarly and pious persons, worthy of all respect, favour us with allocutions upon the sadness of the antagonism of science to their mediæval way of thinking, which betray an ignorance of the first principles of scientific investigation, an incapacity for understanding what a man of science means by veracity, and an unconsciousness of the weight of established scientific truths, which is almost comical.

There is no great force in the *tu quoque* argument, or else the advocates of scientific education might fairly enough retort upon the modern Humanists that they may be learned specialists, but that they possess no such sound foundation for a criticism of life as deserves the name of culture. And, indeed, if we were disposed to be cruel, we might urge that the Humanists have brought this reproach upon themselves, not because they are too full of the spirit of the ancient Greek, but because they lack it.

The period of the Renascence is commonly called that of the "Revival of Letters," as if the influences then brought to bear upon the mind of Western Europe had been wholly exhausted in the field of literature. I think it is very commonly forgotten that the revival of science, effected by the same agency, although less conspicuous, was not less momentous.

In fact, the few and scattered students of nature of that day picked up the clue to her secrets exactly as it fell from the hands of the Greeks a thousand years before. The foundations of mathematics were so well laid by them, that our children learn their geometry from a book written for the schools of Alexandria two thousand years ago. Modern astronomy is the natural continuation and development of the work of Hipparchus and of Ptolemy; modern physics of that of Democritus and of Archimedes; it was long before modern biological science outgrew the knowledge bequeathed to us by Aristotle, by Theophrastus, and by Galen.

We cannot know all the best thoughts and sayings of the Greeks unless we know what they thought about natural phenomena. We cannot fully apprehend their criticism of life unless we understand the extent to which that criticism was affected by scientific conceptions. We falsely pretend to be the inheritors of their culture, unless we are penetrated, as the best minds among them were, with an unhesitating faith that the free employment of reason, in accordance with scientific method, is the sole method of reaching truth.

Thus I venture to think that the pretensions of our modern Humanists to the possession of the monopoly of culture and to the exclusive inheritance of the spirit of antiquity must be abated, if not abandoned. But I should be very sorry that anything I have said should be taken to imply a desire on my part to depreciate the value of classical education, as it might be and as it sometimes is. The native capacities of mankind vary no less than their opportunities; and while culture is one, the road by which one man may best reach it is widely different from that which is most advantageous to another. Again, while scientific education is yet inchoate and tentative, classical education is thoroughly well organised upon the practical experience of generations of teachers. So that, given ample time for learning and destination for ordinary life, or for a literary career, I do not think that a young Englishman in search of culture can do better than follow the course usually marked out for him, supplementing its deficiencies by his own efforts.

But for those who mean to make science their serious occupation; or who intend to follow the profession of medicine; or who have to enter early upon the business of life; for all these, in my opinion, classical education is a mistake; and it is for this reason that I am glad to see "mere literary education and instruction" shut out from the curriculum of Sir Josiah Mason's College, seeing that its inclusion would probably lead to the introduction of the ordinary smattering of Latin and Greek.

Nevertheless, I am the last person to question the importance of genuine literary education, or to suppose that intellectual culture can be complete without it. An exclusively scientific training will bring about a mental twist as surely as an exclusively literary training. The value of the cargo does not compensate for a ship's being out of trim; and I should be very sorry to think that the Scientific College would turn out none but lop-sided men.

There is no need, however, that such a catastrophe

should happen. Instruction in English, French, and German is provided, and thus the three greatest literatures of the modern world are made accessible to the student.

French and German, and especially the latter language, are absolutely indispensable to those who desire full knowledge in any department of science. But even supposing that the knowledge of these languages acquired is not more than sufficient for purely scientific purposes, every Englishman has, in his native tongue, an almost perfect instrument of literary expression; and, in his own literature, models of every kind of literary excellence. If an Englishman cannot get literary culture out of his Bible, his Shakspeare, his Milton, neither, in my belief, will the profoundest study of Homer and Sophocles, Virgil and Horace, give it to him.

Thus, since the constitution of the College makes sufficient provision for literary as well as for scientific education, and since artistic instruction is also contemplated, it seems to me that a fairly complete culture is offered to all who are willing to take advantage of it.

But I am not sure that at this point the "practical" man, scotched but not slain, may ask what all this talk about culture has to do with an Institution, the object of which is defined to be "to promote the prosperity of the manufactures and the industry of the country." He may suggest that what is wanted for this end is not culture, nor even a purely scientific discipline, but simply a knowledge of applied science.

I often wish that this phrase, "applied science," had never been invented. For it suggests that there is a sort of scientific knowledge of direct practical use, which can be studied apart from another sort of scientific knowledge, which is of no practical utility, and which is termed "pure science." But there is no more complete fallacy than this. What people call applied science is nothing but the application of pure science to particular classes of problems. It consists of deductions from those general principles, established by reasoning and observation, which constitute pure science. No one can safely make these deductions until he has a firm grasp of the principles; and he can obtain that grasp only by personal experience of the operations of observation and of reasoning on which they are founded.

Almost all the processes employed in the arts and manufactures fall within the range either of physics or of chemistry. In order to improve them, one must thoroughly understand them; and no one has a chance of really understanding them, unless he has obtained that mastery of principles and that habit of dealing with facts, which is given by long-continued and well-directed purely scientific training in the physical and the chemical laboratory. So that there really is no question as to the necessity of purely scientific discipline, even if the work of the College were limited by the narrowest interpretation of its stated aims.

And, as to the desirableness of a wider culture than that yielded by science alone, it is to be recollected that the improvement of manufacturing processes is only one of the conditions which contribute to the prosperity of industry. Industry is a means and not an end; and mankind work only to get something which they want. What that something is depends partly on their innate, and partly on their acquired, desires.

If the wealth resulting from prosperous industry is to be spent upon the gratification of unworthy desires, if the increasing perfection of manufacturing processes is to be accompanied by an increasing debasement of those who carry them on, I do not see the good of industry and prosperity.

Now it is perfectly true that men's views of what is desirable depend upon their characters; and that the innate proclivities to which we give that name are not touched by any amount of instruction. But it does not follow that even mere intellectual education may not, to an indefinite extent, modify the practical manifestation of the characters of men in their actions, by supplying them with motives unknown to the ignorant. A pleasure-loving character will have pleasure of some sort; but, if you give him the choice, he may prefer pleasures which do not degrade him to those which do. And this choice is offered to every man, who possesses in literary or artistic culture a never-failing source of pleasures, which are neither withered by age, nor staled by custom, nor embittered in the recollection by the pangs of self-reproach.

If the Institution opened to-day fulfils the intention of its founder, the picked intelligences among all classes of the population of this district will pass through it. No child born in Birmingham, henceforward, if he have the capacity to profit by the opportunities offered to him, first in the primary and other schools, and afterwards in the Scientific College, need fail to obtain, not merely the instruction, but the culture most appropriate to the conditions of his life.

Within these walls, the future employer and the future artisan may sojourn together for a while, and carry, through all their lives, the stamp of the influences then brought to bear upon them. Hence, it is not beside the mark to remind you, that the prosperity of industry depends not merely upon the improvement of manufacturing processes, not merely upon the ennobling of the individual character, but upon a third condition, namely, a clear understanding of the conditions of social life on the part of both the capitalist and the operative, and their agreement upon common principles of social action. They must learn that social phenomena are as much the expression of natural laws as any others; that no social arrangements can be permanent unless they harmonise with the requirements of social statics and dynamics; and that, in the nature of things, there is an arbiter whose decisions execute themselves.

But this knowledge is only to be obtained by the application of the methods of investigation adopted in physical researches to the investigation of the phenomena of society. Hence, I confess, I should like to see one addition made to the excellent scheme of education propounded for the College, in the shape of provision for the teaching of Sociology. For though we are all agreed that party politics are to have no place in the instruction of the College; yet in this country, practically governed as it is now by universal suffrage, every man who does his duty must exercise political functions. And, if the evils which are inseparable from the good of political liberty are to be checked, if the perpetual oscillation of nations between anarchy and despotism is to be replaced by the steady march of self-restraining freedom; it will be because men will gradually bring themselves to deal with political, as they now deal with scientific questions; to be as ashamed of undue haste and partisan prejudice in the one case as in the other; and to believe that the machinery of society is at least as delicate as that of a spinning-jenny, and as little likely to be improved by the meddling of those who have not taken the trouble to master the principles of its action.

In conclusion, I am sure that I make myself the mouthpiece of all present in offering to the venerable founder of the Institution, which now commences its beneficent career, our congratulations on the completion of his work; and in expressing the conviction, that the remotest posterity will point to it as a crucial instance of the wisdom which natural piety leads all men to ascribe to their ancestors.

II.

UNIVERSITIES: ACTUAL AND IDEAL.

ELECTED by the suffrages of your four Nations, Rector of the ancient University of which you are scholars, I take the earliest opportunity which has presented itself since my restoration to health, of delivering the Address which, by long custom, is expected of the holder of my office.

My first duty in opening that Address, is to offer you my most hearty thanks for the signal honour you have conferred upon me—an honour of which, as a man unconnected with you by personal or by national ties, devoid of political distinction, and a plebeian who stands by his order, I could not have dreamed. And it was the more surprising to me, as the five-and-twenty years which have passed over my head since I reached intellectual manhood, have been largely spent in no half-hearted advocacy of doctrines which have not yet found favour in the eyes of Academic respectability; so that, when the proposal to nominate me for your Rector came, I was almost as much astonished as was Hal o' the Wynd, "who fought for his own hand," by the Black Douglas's proffer of knighthood. And I fear that my acceptance must be taken as evidence that, less wise than the Armourer of Perth, I have not yet done with soldiering.

In fact, if, for a moment, I imagined that your intention was simply, in the kindness of your hearts, to do me honour; and that the Rector of your University, like that of some other Universities, was one of those happy beings who sit in glory for three years, with nothing to do for it save the making of a speech, a conversation with my distinguished predecessor soon dispelled the dream. I found that, by the constitution of the University of Aberdeen, the incumbent of the Rectorate is, if not a power, at any rate a potential energy; and that, whatever may be his chances of success or failure, it is his duty to convert that potential energy into a living force, directed towards such ends as may seem to him conducive to the welfare of the corporation of which he is the theoretical head.

I need not tell you that your late Lord Rector took this view of his position, and acted upon it with the comprehensive, far-seeing insight into the actual condition and tendencies, not merely of his own, but of other countries, which is his honourable characteristic among statesmen. I have already done my best, and, as long as I hold my office, I shall continue my endeavours, to follow in the path which he trod; to do what in me lies, to bring this University nearer to the ideal—alas, that I should be obliged to say ideal—of all Universities; which, as I conceive, should be places in which thought is free from all fetters; and in which all sources of knowledge, and all aids to learning, should be accessible to all comers, without distinction of creed or country, riches or poverty.

Do not suppose, however, that I am sanguine enough to expect much to come of any poor efforts of mine. If your annals take any notice of my incumbency, I shall probably go down to posterity as the Rector who was always beaten. But if they add, as I think they will, that my defeats became victories in the hands of my successors, I shall be well content.

The scenes are shifting in the great theatre of the world. The act which commenced with the Protestant Reformation is nearly played out, and a wider and a deeper change than that effected three centuries ago—a reformation, or rather a revolution of thought, the extremes of which are represented by the intellectual heirs of John of Leyden and of Ignatius Loyola, rather than by those of Luther and of Leo—is waiting to come on, nay, visible behind the scenes to those who have good eyes. Men are beginning, once more, to awake to the fact that matters of belief and of speculation are of absolutely infinite practical importance; and are drawing off from that sunny country "where it is always afternoon"—the sleepy hollow of broad indifferentism—to range themselves under their natural banners. Change is in the air. It is whirling feather-heads into all sorts of eccentric orbits, and filling the steadiest with a sense of insecurity. It insists on reopening all questions and asking all institutions, however venerable, by what right they exist, and whether they are, or are not, in harmony with the real or supposed wants of mankind. And it is remarkable that these searching inquiries are not so much forced on institutions from without, as developed from within. Consummate scholars question the value of learning; priests contemn dogma; and women turn their backs upon man's ideal of perfect womanhood, and seek satisfaction in apocalyptic visions of some, as yet unrealised, epicene reality.

If there be a type of stability in this world, one would be inclined to look for it in the old Universities of England. But it has been my business of late to hear a good deal about what is going on in these famous corporations; and I have been filled with astonishment by the evidences of internal fermentation which they exhibit. If Gibbon could revisit the ancient seat of learning of which he has written so cavalierly, assuredly he would no longer speak of "the monks of Oxford sunk in prejudice and port." There, as elsewhere, port has gone out of fashion, and so has prejudice—at least that particular fine, old, crusted sort of prejudice to which the great historian alludes.

Indeed, things are moving so fast in Oxford and Cambridge, that, for my part, I rejoiced when the Royal Commission, of which I am a member, had finished and presented the Report which related to these Universities; for we should have looked like mere plagiarists, if, in consequence of a little longer delay in issuing it, all the measures of reform we proposed had been anticipated by the spontaneous action of the Universities themselves.

A month ago I should have gone on to say that

one might speedily expect changes of another kind in Oxford and Cambridge. A Commission has been inquiring into the revenues of the many wealthy societies, in more or less direct connection with the Universities, resident in those towns. It is said that the Commission has reported, and that, for the first time in recorded history, the nation, and perhaps the Colleges themselves, will know what they are worth. And it was announced that a statesman, who, whatever his other merits or defects, has aims above the level of mere party fighting, and a clear vision into the most complex practical problems, meant to deal with these revenues.

But, *Bos locutus est.* That mysterious independent variable of political calculation, Public Opinion—which some whisper is, in the present case, very much the same thing as publican's opinion—has willed otherwise. The Heads may return to their wonted slumbers—at any rate for a space.

Is the spirit of change, which is working thus vigorously in the South, likely to affect the Northern Universities, and if so, to what extent? The violence of fermentation depends, not so much on the quantity of the yeast, as on the composition of the wort, and its richness in fermentable material; and, as a preliminary to the discussion of this question, I venture to call to your minds the essential and fundamental differences between the Scottish and the English type of University.

Do not charge me with anything worse than official egotism, if I say that these differences appear to be largely symbolised by my own existence. There is no Rector in an English University. Now, the organisation of the members of an University into Nations, with their elective Rector, is the last relic of the primitive constitution of Universities. The Rectorate was the most important of all offices in that University of Paris, upon the model of which the University of Aberdeen was fashioned; and which was certainly a great and flourishing institution in the twelfth century.

Enthusiasts for the antiquity of one of the two acknowledged parents of all Universities, indeed, do not hesitate to trace the origin of the "Studium Parisiense" up to that wonderful king of the Franks and Lombards, Karl, surnamed the Great, whom we all called Charlemagne, and believed to be a Frenchman, until a learned historian, by beneficent iteration, taught us better. Karl is said not to have been much of a scholar himself, but he had the wisdom of which knowledge is only the servitor. And that wisdom enabled him to see that ignorance is one of the roots of all evil.

In the Capitulary which enjoins the foundation of monasterial and cathedral schools, he says: "Right action is better than knowledge; but in order to do what is right, we must know what is right."[1] An irrefragable truth, I fancy. Acting upon it, the king took pretty full compulsory powers, and carried into

[1] "Quamvis enim melius sit bene facere quam nosse, prius tamen est nosse quam facere."—"Karoli Magni Regis Constitutio de Scholis per singula Episcopia et Monasteria instituendis," addressed to the Abbot of Fulda. Baluzius, "Capitularia Regum Francorum," T. i., p. 202.

effect a really considerable and effectual scheme of elementary education through the length and breadth of his dominions.

No doubt the idolaters out by the Elbe, in what is now part of Prussia, objected to the Frankish king's measures; no doubt the priests, who had never hesitated about sacrificing all unbelievers in their fantastic deities and futile conjurations, were the loudest in chanting the virtues of toleration; no doubt they denounced as a cruel persecutor the man who would not allow them, however sincere they might be, to go on spreading delusions which debased the intellect, as much as they deadened the moral sense, and undermined the bonds of civil allegiance; no doubt, if they had lived in these times, they would have been able to show, with ease, that the king's proceedings were totally contrary to the best liberal principles. But it may be said, in justification of the Teutonic ruler, first, that he was born before those principles, and did not suspect that the best way of getting disorder into order was to let it alone; and, secondly, that his rough and questionable proceedings did, more or less, bring about the end he had in view. For, in a couple of centuries, the schools he sowed broadcast produced their crop of men, thirsting for knowledge and craving for culture. Such men gravitating towards Paris, as a light amidst the darkness of evil days, from Germany, from Spain, from Britain, and from Scandinavia, came together by natural affinity. By degrees they banded themselves into a society, which, as its end was the knowledge of all things knowable, called itself a "*Studium Generale;*" and when it had grown into a recognised corporation, acquired the name of "*Universitas Studii Generalis,*" which, mark you, means not a "Useful Knowledge Society," but a "Knowledge-of-things-in-general Society."

And thus the first "University," at any rate on this side of the Alps, came into being. Originally it had but one Faculty, that of Arts. Its aim was to be a centre of knowledge and culture; not to be, in any sense, a technical school.

The scholars seem to have studied Grammar, Logic, and Rhetoric; Arithmetic and Geometry; Astronomy; Theology; and Music. Thus, their work, however imperfect and faulty, judged by modern lights, it may have been, brought them face to face with all the leading aspects of the many-sided mind of man. For these studies did really contain, at any rate in embryo—sometimes, it may be, in caricature—what we now call Philosophy, Mathematical and Physical Science, and Art. And I doubt if the curriculum of any modern University shows so clear and generous a comprehension of what is meant by culture, as this old Trivium and Quadrivium does.

The students who had passed through the University course, and had proved themselves competent to teach, became masters and teachers of their younger brethren. Whence the distinction of Masters and Regents on the one hand, and Scholars on the other.

Rapid growth necessitated organisation. The Masters and Scholars of various tongues and countries grouped themselves into four Nations; and the

Nations, by their own votes at first, and subsequently by those of their Procurators, or representatives, elected their supreme head and governor, the Rector —at that time the sole representative of the University, and a very real power, who could defy Provosts interfering from without; or could inflict even corporal punishment on disobedient members within the University.

Such was the primitive constitution of the University of Paris. It is in reference to this original state of things that I have spoken of the Rectorate, and all that appertains to it, as the sole relic of that constitution.

But this original organisation did not last long. Society was not then, any more than it is now, patient of culture, as such. It says to everything, "Be useful to me, or away with you." And to the learned, the unlearned man said then, as he does now, "What is the use of all your learning, unless you can tell me what I want to know? I am here blindly groping about, and constantly damaging myself by collision with three mighty powers, the power of the invisible God, the power of my fellow Man, and the power of brute Nature. Let your learning be turned to the study of these powers, that I may know how I am to comport myself with regard to them." In answer to this demand, some of the Masters of the Faculty of Arts devoted themselves to the study of Theology, some to that of Law, and some to that of Medicine; and they became Doctors—men learned in those technical, or, as we now call them, professional, branches of knowledge. Like cleaving to like, the Doctors formed schools, or Faculties, of Theology, Law, and Medicine, which sometimes assumed airs of superiority over their parent, the Faculty of Arts, though the latter always asserted and maintained its fundamental supremacy.

The Faculties arose by process of natural differentiation out of the primitive University. Other constituents, foreign to its nature, were speedily grafted upon it. One of these extraneous elements was forced into it by the Roman Church, which in those days asserted with effect, that which it now asserts, happily without any effect in these realms, its right of censorship and control over all teaching. The local habitation of the University lay partly in the lands attached to the monastery of S. Geneviève, partly in the diocese of the Bishop of Paris; and he who would teach must have the licence of the Abbot, or of the Bishop, as the nearest representative of the Pope, so to do, which licence was granted by the Chancellors of these Ecclesiastics.

Thus, if I am what archæologists call a "survival" of the primitive head and ruler of the University, your Chancellor stands in the same relation to the Papacy; and, with all respect for his Grace, I think I may say that we both look terribly shrunken when compared with our great originals.

Not so is it with a second foreign element, which silently dropped into the soil of Universities, like the grain of mustard-seed in the parable; and, like that grain, grew into a tree, in whose branches a whole aviary of fowls took shelter. That element is the element of Endowment. It differed from the preceding, in its original design to serve as a prop to the young plant, not to be a parasite upon it. The charitable and the humane, blessed with wealth, were very early penetrated by the misery of the poor student. And the wise saw that intellectual ability is not so common or so unimportant a gift that it should be allowed to run to waste upon mere handicrafts and chares. The man who was a blessing to his contemporaries, but who so often has been converted into a curse, by the blind adherence of his posterity to the letter, rather than to the spirit, of his wishes—I mean the "pious founder"—gave money and lands, that the student, who was rich in brain and poor in all else, might be taken from the plough or from the stithy, and enabled to devote himself to the higher service of mankind; and built colleges and halls in which he might be not only housed and fed, but taught.

The Colleges were very generally placed in strict subordination to the University by their founders; but, in many cases, their endowment, consisting of land, has undergone an "unearned increment," which has given these societies a continually increasing weight and importance as against the unendowed, or fixedly endowed, University. In Pharaoh's dream, the seven lean kine eat up the seven fat ones. In the reality of historical fact, the fat Colleges have eaten up the lean Universities.

Even here in Aberdeen, though the causes at work

may have been somewhat different, the effects have been similar; and you see how much more substantial an entity is the Very Reverend the Principal, analogue, if not homologue, of the Principals of King's College, than the Rector, lineal representative of the ancient monarchs of the University, though now, little more than a "king of shreds and patches."

Do not suppose that, in thus briefly tracing the process of University metamorphosis, I have had any intention of quarrelling with its results. Practically, it seems to me that the broad changes effected in 1858 have given the Scottish Universities a very liberal constitution, with as much real approximation to the primitive state of things as is at all desirable. If your fat kine have eaten the lean, they have not lain down to chew the cud ever since. The Scottish Universities, like the English, have diverged widely enough from their primitive model; but I cannot help thinking that the northern form has remained more faithful to its original, not only in constitution, but, what is more to the purpose, in view of the cry for change, in the practical application of the endowments connected with it.

In Aberdeen, these endowments are numerous, but so small that, taken altogether, they are not equal to the revenue of a single third-rate English college. They are scholarships, not fellowships; aids to do work—not rewards for such work as it lies within the reach of an ordinary, or even an extraordinary, young man to do. You do not think that passing a respectable examination is a fair equivalent for an income, such as many a gray-headed veteran, or clergyman, would envy; and which is larger than the endowment of many Regius chairs. You do not care to make your University a school of manners for the rich; of sports for the athletic; or a hot-bed of high-fed, hypercritical refinement, more destructive to vigour and originality than are starvation and oppression. No; your little Bursaries of ten and twenty (I believe even fifty) pounds a year, enable any boy who has shown ability in the course of his education in those remarkable primary schools, which have made Scotland the power she is, to obtain the highest culture the country can give him; and when he is armed and equipped, his Spartan Alma Mater tells him that, so far, he has had his wages for his work, and that he may go and earn the rest.

When I think of the host of pleasant, monied, well-bred young gentlemen, who do a little learning and much boating by Cam and Isis, the vision is a pleasant one; and, as a patriot, I rejoice that the youth of the upper and richer classes of the nation receive a wholesome and a manly training, however small may be the modicum of knowledge they gather, in the intervals of this, their serious business. I admit, to the full, the social and political value of that training. But, when I proceed to consider that these young men may be said to represent the great bulk of what the Colleges have to show for their enormous wealth, plus, at least, a hundred and fifty pounds a year apiece which each undergraduate costs his parents or guardians, I feel inclined to ask, whether the rate-in-aid of the education of the wealthy and professional classes, thus levied on the resources of the community, is not, after all, a little heavy? And, still further, I am tempted to inquire what has become of the indigent scholars, the sons of the masses of the people whose daily labour just suffices to meet their daily wants, for whose benefit these rich foundations were largely, if not mainly, instituted? It seems as if Pharaoh's dream had been rigorously carried out, and that even the fat scholar has eaten the lean one. And when I turn from this picture to the no less real vision of many a brave and frugal Scotch boy, spending his summer in hard manual labour, that he may have the privilege of wending his way in autumn to this University, with a bag of oatmeal, ten pounds in his pocket, and his own stout heart to depend upon through the northern winter; not bent on seeking

"The bubble reputation at the cannon's mouth,"

but determined to wring knowledge from the hard hands of penury; when I see him win through all such outward obstacles to positions of wide usefulness and well-earned fame; I cannot but think that, in essence, Aberdeen has departed but little from the primitive intention of the founders of Universities, and that the spirit of reform has so much to do on the other side of the Border, that it may be long before he has leisure to look this way.

As compared with other actual Universities, then, Aberdeen, may, perhaps, be well satisfied with itself. But do not think me an impracticable dreamer, if I ask you not to rest and be thankful in this state of satisfaction; if I ask you to consider awhile, how this actual good stands related to that ideal better, towards which both men and institutions must progress, if they would not retrograde.

In an ideal University, as I conceive it, a man should be able to obtain instruction in all forms of knowledge, and discipline in the use of all the methods by which knowledge is obtained. In such an University, the force of living example should fire the student with a noble ambition to emulate the learning of learned men, and to follow in the footsteps of the explorers of new fields of knowledge. And the very air he breathes should be charged with that enthusiasm for truth, that fanaticism of veracity, which is a greater possession than much learning; a nobler gift than the power of increasing knowledge; by so much greater and nobler than these, as the moral nature of man is greater than the intellectual; for veracity is the heart of morality.

But the man who is all morality and intellect, although he may be good and even great, is, after all, only half a man. There is beauty in the moral world and in the intellectual world; but there is also a beauty which is neither moral nor intellectual—the beauty of the world of Art. There are men who are devoid of the power of seeing it, as there are men who are born deaf and blind, and the loss of those, as of these, is simply infinite. There are others in whom it is an overpowering passion; happy men, born with the productive, or at lowest, the appreciative, genius of the

Artist. But, in the mass of mankind, the Æsthetic faculty, like the reasoning power and the moral sense, needs to be roused, directed, and cultivated; and I know not why the development of that side of his nature, through which man has access to a perennial spring of ennobling pleasure, should be omitted from any comprehensive scheme of University education.

All Universities recognise Literature in the sense of the old Rhetoric, which is art incarnate in words. Some, to their credit, recognise Art in its narrower sense, to a certain extent, and confer degrees for proficiency in some of its branches. If there are Doctors of Music, why should there be no Masters of Painting, of Sculpture, of Architecture? I should like to see Professors of the Fine Arts in every University; and instruction in some branch of their work made a part of the Arts curriculum.

I just now expressed the opinion that, in our ideal University, a man should be able to obtain instruction in all forms of knowledge. Now, by "forms of knowledge" I mean the great classes of things knowable; of which the first, in logical, though not in natural, order is knowledge relating to the scope and limits of the mental faculties of man; a form of knowledge which, in its positive aspect, answers pretty much to Logic and part of Psychology, while, on its negative and critical side, it corresponds with Metaphysics.

A second class comprehends all that knowledge which relates to man's welfare, so far as it is determined by his own acts, or what we call his conduct. It answers to Moral and Religious philosophy. Practically, it is the most directly valuable of all forms of knowledge, but speculatively, it is limited and criticised by that which precedes and by that which follows it in my order of enumeration.

A third class embraces knowledge of the phenomena of the Universe, as that which lies about the individual man: and of the rules which those phenomena are observed to follow in the order of their occurrence, which we term the laws of Nature.

This is what ought to be called Natural Science, or Physiology, though those terms are hopelessly diverted from such a meaning; and it includes all exact knowledge of natural fact, whether Mathematical, Physical, Biological, or Social.

Kant has said that the ultimate object of all knowledge is to give replies to these three questions: What can I do? What ought I to do? What may I hope for? The forms of knowledge which I have enumerated, should furnish such replies as are within human reach, to the first and second of these questions. While to the third, perhaps the wisest answer is, "Do what you can to do what you ought, and leave hoping and fearing alone."

If this be a just and an exhaustive classification of the forms of knowledge, no question as to their relative importance, or as to the superiority of one to the other, can be seriously raised.

On the face of the matter, it is absurd to ask whether it is more important to know the limits of one's powers; or the ends for which they ought to be exerted; or the conditions under which they must be exerted. One may as well inquire which of the terms of a Rule of Three sum one ought to know, in order to get a trustworthy result. Practical life is such a sum, in which your duty multiplied into your capacity, and divided by your circumstances, gives you the fourth term in the proportion, which is your deserts, with great accuracy. All agree, I take it, that men ought to have these three kinds of knowledge. The so-called "conflict of studies" turns upon the question of how they may best be obtained.

The founders of Universities held the theory that the Scriptures and Aristotle taken together, the latter being limited by the former, contained all knowledge worth having, and that the business of philosophy was to interpret and co-ordinate these two. I imagine that in the twelfth century this was a very fair conclusion from known facts. Nowhere in the world, in those days, was there such an encyclopædia of knowledge of all three classes, as is to be found in those writings. The scholastic philosophy is a wonderful monument of the patience and ingenuity with which the human mind toiled to build up a logically consistent theory of the Universe, out of such materials. And that philosophy is by no means dead and buried, as many vainly suppose. On the contrary, numbers of men of no mean learning and accomplishment, and sometimes of rare power and subtlety of thought, hold by it as the best theory of things which has yet been stated. And, what is still more remarkable, men who speak the language of modern philosophy, nevertheless think the thoughts of the schoolmen.

"The voice is the voice of Jacob, but the hands are the hands of Esau." Every day I hear "Cause," "Law," "Force," "Vitality," spoken of as entities, by people who can enjoy Swift's joke about the meat-roasting quality of the smoke-jack, and comfort themselves with the reflection that they are not even as those benighted schoolmen.

Well, this great system had its day, and then it was sapped and mined by two influences. The first was the study of classical literature, which familiarised men with methods of philosophising; with conceptions of the highest Good; with ideas of the order of Nature; with notions of Literary and Historical Criticism; and, above all, with visions of Art, of a kind which not only would not fit into the scholastic scheme, but showed them a pre-Christian, and indeed altogether un-Christian world, of such grandeur and beauty that they ceased to think of any other. They were as men who had kissed the Fairy Queen, and wandering with her in the dim loveliness of the underworld, cared not to return to the familiar ways of home and fatherland, though they lay, at arm's length, overhead. Cardinals were more familiar with Virgil than with Isaiah; and Popes laboured, with great success, to re-paganise Rome.

The second influence was the slow, but sure, growth of the physical sciences. It was discovered that some results of speculative thought, of immense practical and theoretical importance, can be verified by observation; and are always true, however severely they may be tested. Here, at any rate, was knowledge, to the certainty of which no authority could add, or take away, one jot or tittle, and to which the tradition of a thousand years was as insignificant as the hearsay of yesterday. To the scholastic system, the study of classical literature might be inconvenient and distracting, but it was possible to hope that it could be kept within bounds. Physical science, on the other hand, was an irreconcilable enemy, to be excluded at all hazards. The College of Cardinals has not distinguished itself in Physics or Physiology; and no Pope has, as yet, set up public laboratories in the Vatican.

People do not always formulate the beliefs on which they act. The instinct of fear and dislike is quicker than the reasoning process; and I suspect that, taken in conjunction with some other causes, such instinctive aversion is at the bottom of the long exclusion of any serious discipline in the physical sciences from the general curriculum of Universities; while, on the other hand, classical literature has been gradually made the backbone of the Arts course.

I am ashamed to repeat here what I have said elsewhere, in season and out of season, respecting the value of Science as knowledge and discipline. But the other day I met with some passages in the Address to another Scottish University, of a great thinker, recently lost to us, which express so fully, and yet so tersely, the truth in this matter, that I am fain to quote them:—

"To question all things;—never to turn away from any difficulty; to accept no doctrine either from ourselves or from other people without a rigid scrutiny by negative criticism; letting no fallacy, or incoherence, or confusion of thought step by unperceived; above all, to insist upon having the meaning of a word clearly understood before using it, and the meaning of a proposition before assenting to it;—these are the lessons we learn" from workers in Science. "With all this vigorous management of the negative element, they inspire no scepticism about the reality of truth or indifference to its pursuit. The noblest enthusiasm, both for the search after truth and for applying it to its highest uses, pervades those writers." "In cultivating, therefore," science as an essential ingredient in education, "we are all the while laying an admirable foundation for ethical and philosophical culture."[1]

The passages I have quoted were uttered by John Stuart Mill; but you cannot hear inverted commas, and it is therefore right that I should add, without delay, that I have taken the liberty of substituting "workers in science" for "ancient dialecticians," and "Science as an essential ingredient in education" for "the ancient languages as our best literary education." Mill did, in fact, deliver a noble panegyric upon classical studies. I do not doubt its justice, nor presume to question its wisdom. But I venture to maintain that no wise or just judge, who has a knowledge of the facts, will hesitate to say that it applies with equal force to scientific training.

[1] Inaugural Address delivered to the University of St. Andrew February 1, 1867, by J. S. Mill, Rector of the University (pp. 32, 33.

But it is only fair to the Scottish Universities to point out that they have long understood the value of Science as a branch of general education. I observe, with the greatest satisfaction, that candidates for the degree of Master of Arts in this University are required to have a knowledge, not only of Mental and Moral Philosophy, and of Mathematics and Natural Philosophy, but of Natural History, in addition to the ordinary Latin and Greek course; and that a candidate may take honours in these subjects and in Chemistry.

I do not know what the requirements of your examiners may be, but I sincerely trust they are not satisfied with a mere book knowledge of these matters. For my own part, I would not raise a finger, if I could thereby introduce mere book work in science into every Arts curriculum in the country. Let those who want to study books devote themselves to Literature, in which we have the perfection of books, both as to substance and as to form. If I may paraphrase Hobbes's well-known aphorism, I would say that "books are the money of Literature, but only the counters of Science," Science (in the sense in which I now use the term) being the knowledge of fact, of which every verbal description is but an incomplete and symbolic expression. And be assured that no teaching of science is worth anything, as a mental discipline, which is not based upon direct perception of the facts, and practical exercise of the observing and logical faculties upon them. Even in such a simple matter as the mere comprehension of form, ask the most practised and widely informed anatomist what is the difference between his knowledge of a structure which he has read about, and his knowledge of the same structure when he has seen it for himself; and he will tell you that the two things are not comparable—the difference is infinite. Thus I am very strongly inclined to agree with some learned schoolmasters who say that, in their experience, the teaching of science is all waste time. As they teach it, I have no doubt it is. But to teach it otherwise, requires an amount of personal labour and a development of means and appliances, which must strike horror and dismay into a man accustomed to mere book work; and who has been in the habit of teaching a class of fifty without much strain upon his energies. And this is one of the real difficulties in the way of the introduction of physical science into the ordinary University course, to which I have alluded. It is a difficulty which will not be overcome, until years of patient study have organised scientific teaching as well as, or I hope better than, classical teaching has been organised hitherto.

A little while ago, I ventured to hint a doubt as to the perfection of some of the arrangements in the ancient Universities of England; but, in their provision for giving instruction in Science as such, and without direct reference to any of its practical applications, they have set a brilliant example. Within the last twenty years, Oxford alone has sunk more than a hundred and twenty thousand pounds in building and furnishing Physical, Chemical, and Physiological Laboratories, and a magnificent Museum, arranged with an almost luxurious regard for the needs of the student. Cambridge, less rich, but aided by the munificence of her Chancellor, is taking the same course; and, in a few years, it will be for no lack of the means and appliances of sound teaching, if the mass of English University men remain in their present state of barbarous ignorance of even the rudiments of scientific culture.

Yet another step needs to be made before Science can be said to have taken its proper place in the Universities. That is its recognition as a Faculty, or branch of study demanding recognition and special organisation, on account of its bearing on the wants of mankind. The Faculties of Theology, Law, and Medicine, are technical schools, intended to equip men who have received general culture, with the special knowledge which is needed for the proper performance of the duties of clergymen, lawyers, and medical practitioners.

When the material well-being of the country depended upon rude pasture and agriculture, and still ruder mining; in the days when all the innumerable applications of the principles of physical science to practical purposes were non-existent even as dreams; days which men living may have heard their fathers speak of; what little physical science could be seen to bear directly upon human life, lay within the province of Medicine. Medicine was the foster-mother of Chemistry, because it has to do with the preparation of drugs and the detection of poisons; of Botany, because it enabled the physician to recognise medicinal herbs; of Comparative Anatomy and Physiology, because the man who studied Human Anatomy and Physiology for purely medical purposes was led to extend his studies to the rest of the animal world.

Within my recollection, the only way in which a student could obtain anything like a training in Physical Science, was by attending the lectures of the Professors of Physical and Natural Science attached to the Medical Schools. But, in the course of the last thirty years, both foster-mother and child have grown so big, that they threaten not only to crush one another, but to press the very life out of the unhappy student who enters the nursery; to the great detriment of all three.

I speak in the presence of those who know practically what medical education is; for I may assume that a large proportion of my hearers are more or less advanced students of medicine. I appeal to the most industrious and conscientious among you, to those who are most deeply penetrated with a sense of the extremely serious responsibilities which attach to the calling of a medical practitioner, when I ask whether, out of the four years which you devote to your studies, you ought to spare even so much as an hour for any work which does not tend directly to fit you for your duties?

Consider what that work is. Its foundation is a sound and practical acquaintance with the structure of the human organism, and with the modes and conditions of its action in health. I say a sound and practical acquaintance, to guard against the supposition that my intention is to suggest that you ought all to be minute anatomists and accomplished physiologists. The devotion of your whole four years to Anatomy and Physiology alone, would be totally insufficient to attain that end. What I mean is, the sort of practical, familiar, finger-end knowledge which a watchmaker has of a watch, and which you expect that craftsman, as an honest man, to have, when you entrust a watch that goes badly, to him. It is a kind of knowledge which is to be acquired, not in the lecture-room, nor in the library, but in the dissecting-room and the laboratory. It is to be had, not by sharing your attention between these and sundry other subjects, but by concentrating your minds, week after week, and month after month, six or seven hours a day, upon all the complexities of organ and function, until each of the greater truths of anatomy and physiology has become an organic part of your minds—until you would know them if you were roused and questioned in the middle of the night, as a man knows the geography of his native place and the daily life of his home. That is the sort of knowledge which, once obtained, is a life-long possession. Other occupations may fill your minds—it may grow dim, and seem to be forgotten—but there it is, like the inscription on a battered and defaced coin, which comes out when you warm it.

If I had the power to remodel Medical Education, the first two years of the medical curriculum should be devoted to nothing but such thorough study of Anatomy and Physiology, with Physiological Chemistry and Physics; the student should then pass a real, practical examination in these subjects; and, having gone through that ordeal satisfactorily, he should be troubled no more with them. His whole mind should then be given with equal intentness, to Therapeutics, in its broadest sense, to Practical Medicine and to Surgery, with instruction in Hygiene and in Medical Jurisprudence; and of these subjects only—surely there are enough of them—should he be required to show a knowledge in his final examination.

I cannot claim any special property in this theory of what the medical curriculum should be, for I find that views, more or less closely approximating these, are held by all who have seriously considered the very grave and pressing question of Medical Reform; and have, indeed, been carried into practice, to some extent, by the most enlightened Examining Boards. I have heard but two kinds of objections to them. There is, first, the objection of vested interests, which I will not deal with here, because I want to make myself as pleasant as I can, and no discussions are so unpleasant as those which turn on such points. And there is, secondly, the much more respectable objection, which takes the general form of the reproach that, in thus limiting the curriculum, we are seeking to narrow it. We are told that the medical man ought to be a person of good education and general information, if his profession is to hold its own among other professions; that he ought to know Botany, or else, if he goes abroad, he will not be able to tell poisonous fruits from edible ones; that he ought to know drugs, as a druggist knows them, or he will not be able to tell sham bark and senna from the real articles; that he ought to know Zoology, because—well, I really have never been able to learn exactly why he is to be expected to know zoology. There is, indeed, a popular superstition, that doctors know all about things that are queer or nasty to the general mind, and may, therefore, be reasonably expected to know the "barbarous binomials" applicable to snakes, snails, and slugs; an amount of information with which the general mind is usually completely satisfied. And there is a scientific superstition that Physiology is largely aided by Comparative Anatomy—a superstition which, like most superstitions, once had a grain of truth at bottom; but the grain has become homœopathic, since Physiology took its modern experimental development, and became what it is now, the application of the principles of Physics and Chemistry to the elucidation of the phenomena of life.

I hold as strongly as any one can do, that the medical practitioner ought to be a person of education and good general culture; but I also hold by the old theory of a Faculty, that a man should have his general culture before he devotes himself to the special studies of that Faculty; and I venture to maintain, that, if the general culture obtained in the Faculty of Arts were what it ought to be, the student would have quite as much knowledge of the fundamental principles of Physics, of Chemistry, and of Biology, as he needs, before he commenced his special medical studies.

Moreover, I would urge, that a thorough study of Human Physiology is, in itself, an education broader and more comprehensive than much that passes under that name. There is no side of the intellect which it does not call into play, no region of human knowledge into which either its roots, or its branches, do not extend; like the Atlantic between the Old and the New Worlds, its waves wash the shores of the two worlds of matter and of mind; its tributary streams flow from both; through its waters, as yet unfurrowed by the keel of any Columbus, lies the road, if such there be, from the one to the other; far away from that North-west Passage of mere speculation, in which so many brave souls have been hopelessly frozen up.

But whether I am right or wrong about all this, the patent fact of the limitation of time remains. As the song runs:—

> "If a man could be sure
> That his life would endure
> For the space of a thousand long years——"

he might do a number of things not practicable under present conditions. Methuselah might, with much propriety, have taken half a century to get his doctor's degree; and might, very fairly, have been required to pass a practical examination upon the contents of the British Museum, before commencing practice as a promising young fellow of two hundred, or thereabouts. But you have four years to do your work in, and are turned loose, to save or slay, at two or three and twenty.

Now, I put it to you, whether you think that,

when you come down to the realities of life—when you stand by the sick-bed, racking your brains for the principles which shall furnish you with the means of interpreting symptoms, and forming a rational theory of the condition of your patient, it will be satisfactory for you to find that those principles are not there—although, to use the examination slang which is unfortunately too familiar to me, you can quite easily "give an account of the leading peculiarities of the *Marsupialia*," or "enumerate the chief characters of the *Compositæ*," or "state the class and order of the animal from which Castoreum is obtained."

I really do not think that state of things will be satisfactory to you; I am very sure it will not be so to your patient. Indeed, I am so narrow-minded myself, that if I had to choose between two physicians—one who did not know whether a whale is a fish or not, and could not tell gentian from ginger, but did understand the applications of the institutes of medicine to his art; while the other, like Talleyrand's doctor, "knew everything, even a little physic"—with all my love for breadth of culture, I should assuredly consult the former.

It is not pleasant to incur the suspicion of an inclination to injure or depreciate particular branches of knowledge. But the fact that one of those which I should have no hesitation in excluding from the medical curriculum, is that to which my own life has been specially devoted, should, at any rate, defend me from the suspicion of being urged to this course by any but the very gravest considerations of the public welfare.

And I should like, further, to call your attention to the important circumstance that, in thus proposing the exclusion of the study of such branches of knowledge as Zoology and Botany, from those compulsory upon the medical student, I am not, for a moment, suggesting their exclusion from the University. I think that sound and practical instruction in the elementary facts and broad principles of Biology should form part of the Arts Curriculum: and here, happily, my theory is in entire accordance with your practice. Moreover, as I have already said, I have no sort of doubt that, in view of the relation of Physical Science to the practical life of the present day, it has the same right as Theology, Law, and Medicine, to a Faculty of its own in which men shall be trained to be professional men of science. It may be doubted whether Universities are the places for technical schools of Engineering, or Applied Chemistry, or Agriculture. But there can surely be little question, that instruction in the branches of Science which lie at the foundation of these Arts, of a far more advanced and special character than could, with any propriety, be included in the ordinary Arts Curriculum, ought to be obtainable by means of a duly organised Faculty of Science in every University.

The establishment of such a Faculty would have the additional advantage of providing, in some measure, for one of the greatest wants of our time and country. I mean the proper support and encouragement of original research.

The other day, an emphatic friend of mine committed himself to the opinion that, in England, it is better for a man's worldly prospects to be a drunkard, than to be smitten with the divine dipsomania of the original investigator. I am inclined to think he was not far wrong. And, be it observed, that the question is not, whether such a man shall be able to make as much out of his abilities as his brother, of like ability, who goes into Law, or Engineering, or Commerce; it is not a question of "maintaining a due number of saddle horses," as George Eliot somewhere puts it—it is a question of living or starving.

If a student of my own subject shows power and originality, I dare not advise him to adopt a scientific career; for, supposing he is able to maintain himself until he has attained distinction, I cannot give him the assurance that any amount of proficiency in the Biological Sciences will be convertible into, even the most modest, bread and cheese. And I believe that the case is as bad, or perhaps worse, with other branches of Science. In this respect Britain, whose immense wealth and prosperity hang upon the thread of Applied Science, is far behind France, and infinitely behind Germany.

And the worst of it is, that it is very difficult to see one's way to any immediate remedy for this state of affairs which shall be free from a tendency to become worse than the disease.

Great schemes for the Endowment of Research

have been proposed. It has been suggested, that Laboratories for all branches of Physical Science, provided with every apparatus needed by the investigator, shall be established by the State: and shall be accessible, under due conditions and regulations, to all properly qualified persons. I see no objection to the principle of such a proposal. If it be legitimate to spend great sums of money on public Libraries and public collections of Painting and Sculpture, in aid of the man of letters, or the Artist, or for the mere sake of affording pleasure to the general public, I apprehend that it cannot be illegitimate to do as much for the promotion of scientific investigation. To take the lowest ground, as a mere investment of money, the latter is likely to be much more immediately profitable. To my mind, the difficulty in the way of such schemes is not theoretical, but practical. Given the laboratories, how are the investigators to be maintained? What career is open to those who have been thus encouraged to leave bread-winning pursuits? If they are to be provided for by endowment, we come back to the College Fellowship system, the results of which, for Literature, have not been so brilliant that one would wish to see it extended to Science; unless some much better securities, than at present exist, can be taken that it will foster real work. You know that among the Bees, it depends on the kind of cell in which the egg is deposited, and the quantity and quality of food which is supplied to the grub, whether it shall turn out a busy little worker or a big idle queen. And, in the human hive, the cells of the endowed larvæ are always tending to enlarge, and their food to improve, until we get queens, beautiful to behold, but which gather no honey and build no comb.

I do not say that these difficulties may not be overcome, but their gravity is not to be lightly estimated.

In the meanwhile, there is one step in the direction of the endowment of research which is free from such objections. It is possible to place the scientific inquirer in a position in which he shall have ample leisure and opportunity for original work, and yet shall give a fair and tangible equivalent for those privileges. The establishment of a Faculty of Science in every University, implies that of a corresponding number of Professorial chairs, the incumbents of which need not be so burdened with teaching as to deprive them of ample leisure for original work. I do not think that it is any impediment to an original investigator to have to devote a moderate portion of his time to lecturing, or superintending practical instruction. On the contrary, I think it may be, and often is, a benefit to be obliged to take a comprehensive survey of your subject; or to bring your results to a point, and give them, as it were, a tangible objective existence. The besetting sins of the investigator are two: the one is the desire to put aside a subject, the general bearings of which he has mastered himself, and pass on to something which has the attraction of novelty; and the other, the desire for too much perfection, which leads him to

"Add and alter many times,
Till all be ripe and rotten;"

to spend the energies which should be reserved for action, in whitening the decks and polishing the guns.

The obligation to produce results for the instruction of others, seems to me to be a more effectual check on these tendencies, than even the love of usefulness or the ambition for fame.

But supposing the Professorial forces of our University to be duly organised, there remains an important question, relating to the teaching power, to be considered. Is the Professorial system—the system, I mean, of teaching in the lecture-room alone, and leaving the student to find his own way when he is outside the lecture-room—adequate to the wants of learners? In answering this question, I confine myself to my own province, and I venture to reply for Physical Science, assuredly and undoubtedly, No. As I have already intimated, practical work in the Laboratory is absolutely indispensable, and that practical work must be guided and superintended by a sufficient staff of Demonstrators, who are for Science what Tutors are for other branches of study. And there must be a good supply of such Demonstrators. I doubt if the practical work of more than twenty students can be properly superintended by one Demonstrator. If we take the working day at six hours, that is less than twenty minutes apiece—not a very large allowance of time for helping a dull man, for correcting an inaccurate one, or even for making an intelligent student clearly apprehend what he is about. And, no doubt, the supplying of a proper amount of this tutorial, practical teaching, is a difficulty in the way of giving proper instruction in Physical Science in such Universities as that of Aberdeen, which are devoid of endowments; and, unlike the English Universities, have no moral claim on the funds of richly endowed bodies to supply their wants.

Examination—thorough, searching examination—is an indispensable accompaniment of teaching; but I am almost inclined to commit myself to the very heterodox proposition that it is a necessary evil. I am a very old Examiner, having, for some twenty years past, been occupied with examinations on a considerable scale, of all sorts and conditions of men, and women too,—from the boys and girls of elementary schools to the candidates for Honours and Fellowships in the Universities. I will not say that, in this case as in so many others, the adage, that familiarity breeds contempt, holds good; but my admiration for the existing system of examination and its products, does not wax warmer as I see more of it. Examination, like fire, is a good servant, but a bad master; and there seems to me to be some danger of its becoming our master. I by no means stand alone in this opinion. Experienced friends of mine do not hesitate to say that students whose career they watch, appear to them to become deteriorated by the constant effort to pass this or that examination, just as we hear of men's brains becoming affected by the daily necessity of catching a train. They work to pass, not to know; and outraged Science takes her revenge. They do pass, and they don't know. I have passed sundry examinations in my time, not without credit, and I confess I am ashamed to think how very little real knowledge underlay the torrent of stuff which I was able to pour out on paper. In fact, that which examination, as ordinarily conducted, tests, is simply a man's power of work under stimulus, and his capacity for rapidly and clearly producing that which, for the time, he has got into his mind. Now, these faculties are by no means to be despised. They are of great value in practical life, and are the making of many an advocate, and of many a so-called statesman. But in the pursuit of truth, scientific or other, they count for very little, unless they are supplemented by that long-continued, patient "intending of the mind," as Newton phrased it, which makes very little show in Examinations. I imagine that an Examiner who knows his students personally, must not unfrequently have found himself in the position of finding A's paper better than B's, though his own judgment tells him, quite clearly, that B is the man who has the larger share of genuine capacity.

Again, there is a fallacy about Examiners. It is commonly supposed that any one who knows a subject is competent to teach it; and no one seems to doubt that any one who knows a subject is competent to examine in it. I believe both these opinions to be serious mistakes: the latter, perhaps, the more serious of the two. In the first place, I do not believe that any one who is not, or has not been, a teacher is really qualified to examine advanced students. And in the second place, Examination is an Art, and a difficult one, which has to be learned like all other arts.

Beginners always set too difficult questions—partly because they are afraid of being suspected of ignorance if they set easy ones, and partly from not understanding their business. Suppose that you want to test the relative physical strength of a score of young men. You do not put a hundredweight down before them, and tell each to swing it round. If you do, half of them won't be able to lift it at all, and only one or two will be able to perform the task. You must give them half a hundredweight, and see how they manœuvre that, if you want to form any estimate of the muscular strength of each. So, a practised Examiner will seek for information respecting the mental vigour and training of candidates from the way in which they deal with questions easy enough to let reason, memory, and method have free play.

No doubt, a great deal is to be done by the careful selection of Examiners, and by the copious introduction of practical work, to remove the evils inseparable from examination; but, under the best of circumstances, I believe that examination will remain but an imperfect test of knowledge, and a still more imperfect test of capacity, while it tells next to nothing about a man's power as an investigator.

There is much to be said in favour of restricting the highest degrees in each Faculty, to those who have shown evidence of such original power, by prosecuting a research under the eye of the Professor in whose province it lies; or, at any rate, under conditions which shall afford satisfactory proof that the work is theirs. The notion may sound revolutionary, but it is really very old; for, I take it, that it lies at the bottom of that presentation of a thesis by the candidate for a doctorate, which has now, too often, become little better than a matter of form.

Thus far, I have endeavoured to lay before you, in a too brief and imperfect manner, my views respecting the teaching half—the Magistri and Regentes—of the University of the Future. Now let me turn to the learning half—the Scholares.

If the Universities are to be the sanctuaries of the highest culture of the country, those who would enter that sanctuary, must not come with unwashed hands. If the good seed is to yield its hundredfold harvest, it must not be scattered amidst the stones of ignorance, or the tares of undisciplined indolence and wantonness. On the contrary, the soil must have been carefully prepared, and the Professor should find that the operations of clod-crushing, draining, and weeding, and even a good deal of planting, have been done by the Schoolmaster.

That is exactly what the Professor does not find in any University in the three Kingdoms that I can hear of—the reason of which state of things lies in the extremely faulty organisation of the majority of secondary Schools. Students come to the Universities ill-prepared in classics and mathematics, not at all prepared in anything else; and half their time is spent in learning that which they ought to have known when they came.

I sometimes hear it said that the Scottish Universities differ from the English, in being to a much greater extent places of comparatively elementary education for a younger class of students. But it would seem doubtful if any great difference of this kind really exists; for a high authority, himself Head of an English College, has solemnly affirmed that: "Elementary teaching of youths under twenty is now the only function performed by the University;" and that Colleges are "boarding schools in which the elements of the learned languages are taught to youths."[1]

This is not the first time that I have quoted those remarkable assertions. I should like to engrave them in public view, for they have not been refuted; and I am convinced that if their import is once clearly apprehended, they will play no mean part when the question of University reorganisation, with a view to practical measures, comes on for discussion. You are not responsible for this anomalous state of affairs now; but, as you pass into active life and acquire the political influence to which your education and your position should entitle you, you will become responsible for it, unless each in his sphere does his best to alter it, by insisting on the improvement of secondary Schools.

Your present responsibility is of another, though

[1] "Suggestions for Academical Organisation, with Especial Reference to Oxford." By the Rector of Lincoln.

not less serious, kind. Institutions do not make men, any more than organisation makes life; and even the ideal University we have been dreaming about will be but a superior piece of mechanism, unless each student strive after the ideal of the Scholar. And that ideal, it seems to me, has never been better embodied than by the great Poet, who, though lapped in luxury, the favourite of a Court, and the idol of his countrymen, remained through all the length of his honoured years a Scholar in Art, in Science, and in Life.

> "Would'st shape a noble life? Then cast
> No backward glances towards the past:
> And though somewhat be lost and gone,
> Yet do thou act as one new-born.
> What each day needs, that shalt thou ask;
> Each day will set its proper task.
> Give other's work just share of praise;
> Not of thine own the merits raise.
> Beware no fellow man thou hate:
> And so in God's hands leave thy fate." [1]

---

[1] Goethe, *Zahme Xenien, Vierte Abtheilung.* I should be glad to take credit for the close and vigorous English version; but it is my wife's, and not mine.

III.

TECHNICAL EDUCATION.

Any candid observer of the phenomena of modern society will readily admit that bores must be classed among the enemies of the human race; and a little consideration will probably lead him to the further admission, that no species of that extensive genus of noxious creatures is more objectionable than the educational bore. Convinced as I am of the truth of this great social generalisation, it is not without a certain trepidation that I venture to address you on an educational topic. For, in the course of the last ten years, to go back no farther, I am afraid to say how often I have ventured to speak of education, from that given in the primary schools to that which is to be had in the universities and medical colleges; indeed, the only part of this wide region into which, as yet, I have not adventured is that into which I propose to intrude to-day.

Thus, I cannot but be aware that I am dangerously near becoming the thing which all men fear and fly. But I have deliberately elected to run the risk. For when you did me the honour to ask me to address you, an unexpected circumstance had led me to occupy myself seriously with the question of technical education; and I had acquired the conviction that there are few subjects respecting which it is more important for all classes of the community to have clear and just ideas than this; while, certainly, there is none which is more deserving of attention by the Working Men's Club and Institute Union.

It is not for me to express an opinion whether the considerations, which I am about to submit to you, will be proved by experience to be just or not; but I will do my best to make them clear. Among the many good things to be found in Lord Bacon's works, none is more full of wisdom than the saying that "truth more easily comes out of error than out of confusion." Clear and consecutive wrong-thinking is the next best thing to right-thinking; so that, if I succeed in clearing your ideas on this topic, I shall have wasted neither your time nor my own.

"Technical education," in the sense in which the term is ordinarily used, and in which I am now employing it, means that sort of education which is specially adapted to the needs of men whose business in life it is to pursue some kind of handicraft; it is, in fact, a fine Greco-Latin equivalent for what in good vernacular English would be called "the teaching of handicrafts." And probably, at this stage of our progress, it may occur to many of you to think of the story of the cobbler and his last, and to say to yourselves, though you will be too polite to put the question openly to me, What does the speaker know practically about this matter? What is his handicraft? I think the question is a very proper one, and unless I were prepared to answer it, I hope satisfactorily, I should have chosen some other theme.

The fact is, I am, and have been, any time these thirty years, a man who works with his hands—a handicraftsman. I do not say this in the broadly metaphorical sense in which fine gentlemen, with all the delicacy of Agag about them, trip to the hustings about election time, and protest that they too are working men. I really mean my words to be taken in their direct, literal, and straightforward sense. In fact, if the most nimble-fingered watchmaker among you will come to my workshop, he may set me to put a watch together, and I will set him to dissect, say, a blackbeetle's nerves. I do not wish to vaunt, but I am inclined to think that I shall manage my job to his satisfaction sooner than he will do his piece of work to mine.

In truth, anatomy, which is my handicraft, is one of the most difficult kinds of mechanical labour, involving, as it does, not only lightness and dexterity of hand, but sharp eyes and endless patience. And you must not suppose that my particular branch of science is especially distinguished for the demand it makes upon skill in manipulation. A similar requirement is made upon all students of physical science. The astronomer, the electrician, the chemist, the mineralogist, the botanist, are constantly called upon to perform manual operations of exceeding delicacy. The progress of all branches of physical science depends upon observation, or on that artificial observation which is termed experiment, of one kind or another; and, the farther we advance, the more practical difficulties surround the investigation of the conditions of the problems offered to us; so that mobile and yet steady hands, guided by clear vision, are more and more in request in the workshops of science.

Indeed, it has struck me that one of the grounds of that sympathy between the handicraftsmen of this country and the men of science, by which it has so often been my good fortune to profit, may, perhaps, lie here. You feel and we feel that, among the so-called learned folks, we alone are brought into contact with tangible facts in the way that you are. You know well enough that it is one thing to write a history of chairs in general, or to address a poem to a throne, or to speculate about the occult powers of the chair of St. Peter; and quite another thing to make with your own hands a veritable chair, that will stand fair and square, and afford a safe and satisfactory resting-place to a frame of sensitiveness and solidity.

So it is with us, when we look out from our scientific handicrafts upon the doings of our learned brethren, whose work is untrammelled by anything "base and mechanical," as handicrafts used to be called when the world was younger, and, in some respects, less wise than now. We take the greatest interest in their pursuits; we are edified by their histories and are charmed with their poems, which sometimes illustrate so remarkably the powers of man's imagination; some of us admire and even humbly try to follow them in their high philosophical excursions, though we know the risk of being snubbed by the inquiry whether grovelling dissectors of monkeys and blackbeetles can hope to enter into the empyreal kingdom of speculation. But still we feel that our business is different; humbler if you will, though the diminution of dignity is, perhaps, compensated by the increase of reality; and that we, like you, have to get our work done in a region where little avails, if the power of dealing with practical tangible facts is wanting. You know that clever talk touching joinery will not make a chair; and I know that it is of about as much value in the physical sciences. Mother Nature is serenely obdurate to honeyed words; only those who understand the ways of things, and can silently and effectually handle them, get any good out of her.

And now, having, as I hope, justified my assumption of a place among handicraftsmen, and put myself right with you as to my qualification, from practical knowledge, to speak about technical education, I will proceed to lay before you the results of my experience as a teacher of a handicraft, and tell you what sort of education I should think best adapted for a boy whom one wanted to make a professional anatomist.

I should say, in the first place, let him have a good English elementary education. I do not mean that he shall be able to pass in such and such a standard —that may or may not be an equivalent expression —but that his teaching shall have been such as to have given him command of the common implements of learning and to have created a desire for the things of the understanding.

Further, I should like him to know the elements of physical science, and especially of physics and chemistry, and I should take care that this elementary knowledge was real. I should like my aspirant to be able to read a scientific treatise in Latin, French, or German, because an enormous amount of anatomical knowledge is locked up in those languages. And especially, I should require some ability to draw—I do not mean artistically, for that is a gift which may be cultivated but cannot be learned, but with fair accuracy. I will not say that everybody can learn even this; for the negative development of the faculty of drawing in some people is almost miraculous. Still everybody, or almost everybody, can learn to write; and, as writing is a kind of drawing, I suppose that the majority of the people who say they cannot draw, and give copious evidence of the accuracy of their assertion, could draw, after a fashion, if they tried. And that "after a fashion" would be better than nothing for my purposes.

Above all things, let my imaginary pupil have preserved the freshness and vigour of youth in his mind as well as his body. The educational abomination of desolation of the present day is the stimulation of young people to work at high pressure by incessant competitive examinations. Some wise man (who probably was not an early riser) has said of early risers in general, that they are conceited all the forenoon and stupid all the afternoon. Now whether this is true of early risers in the common acceptation of the word or not, I will not pretend to say; but it is too often true of the unhappy children who are forced to rise too early in their classes. They are conceited all the forenoon of life, and stupid all its afternoon. The vigour and freshness, which should have been stored up for the purposes of the hard struggle for existence in practical life, have been washed out of them by precocious mental debauchery—by book gluttony and lesson bibbing. Their faculties are worn out by the strain put upon their callow brains, and they are demoralised by worthless childish triumphs before the real work of life begins. I have no compassion for sloth, but youth has more need for intellectual rest than age; and the cheerfulness, the tenacity of purpose, the power of work which make many a successful man what he is, must often be placed to the credit, not of his hours of industry, but to that of his hours of idleness, in boyhood. Even the hardest worker of us all, if he has to deal with anything above mere details, will do well, now and again, to let his brain lie fallow for a space. The next crop of thought will certainly be all the fuller in the ear and the weeds fewer.

This is the sort of education which I should like any one who was going to devote himself to my handicraft to undergo. As to knowing anything about anatomy itself, on the whole I would rather he left that alone until he took it up seriously in my laboratory. It is hard work enough to teach, and I should not like to have superadded to that the possible need of unteaching.

Well, but, you will say, this is Hamlet with the Prince of Denmark left out; your "technical education" is simply a good education, with more attention to physical science, to drawing, and to modern languages, than is common, and there is nothing specially technical about it.

Exactly so; that remark takes us straight to the heart of what I have to say; which is, that, in my judgment, the preparatory education of the handicraftsman ought to have nothing of what is ordinarily understood by "technical" about it.

The workshop is the only real school for a handicraft. The education which precedes that of the workshop should be entirely devoted to the strengthening of the body, the elevation of the moral faculties, and the cultivation of the intelligence; and, especially, to the imbuing the mind with a broad and clear view of the laws of that natural world with the components of which the handicraftsman will have to deal. And, the earlier the period of life at which the handicraftsman has to enter into actual practice of his craft, the more important is it that he should devote the precious hours of preliminary education to things of the mind, which have no direct and immediate bearing on his branch of industry, though they lie at the foundation of all realities.

Now let me apply the lessons I have learned from my handicraft to yours. If any of you were obliged to take an apprentice, I suppose you would like to get a good healthy lad, ready and willing to learn, handy, and with his fingers not all thumbs, as the saying goes. You would like that he should read, write, and cipher well; and, if you were an intelligent master, and your trade involved the application of scientific principles, as so many trades do, you would like him to know enough of the elementary principles of science to understand what was going on. I suppose that, in nine trades out of ten, it would be useful if he could draw; and many of you must have lamented your inability to find out for yourselves what foreigners are doing or have done. So that some knowledge of French and German might, in many cases, be very desirable.

So it appears to me that what you want is pretty much what I want; and the practical question is, How you are to get what you need, under the actual limitations and conditions of life of handicraftsmen in this country?

I think I shall have the assent both of the employers of labour and of the employed as to one of these limitations; which is, that no scheme of technical education is likely to be seriously entertained which will delay the entrance of boys into working life, or prevent them from contributing towards their own support, as early as they do at present. Not only do I believe that any such scheme could not be carried out, but I doubt its desirableness, even if it were practicable.

The period between childhood and manhood is full

of difficulties and dangers, under the most favourable circumstances; and, even among the well-to-do, who can afford to surround their children with the most favourable conditions, examples of a career ruined, before it has well begun, are but too frequent. Moreover, those who have to live by labour must be shaped to labour early. The colt that is left at grass too long makes but a sorry draught-horse, though his way of life does not bring him within the reach of artificial temptations. Perhaps the most valuable result of all education is the ability to make yourself do the thing you have to do, when it ought to be done, whether you like it or not; it is the first lesson that ought to be learned; and, however early a man's training begins, it is probably the last lesson that he learns thoroughly.

There is another reason, to which I have already adverted, and which I would reiterate, why any extension of the time devoted to ordinary school-work is undesirable. In the newly awakened zeal for education, we run some risk of forgetting the truth that while under-instruction is a bad thing, over-instruction may possibly be a worse.

Success in any kind of practical life is not dependent solely, or indeed chiefly, upon knowledge. Even in the learned professions, knowledge, alone, is of less consequence than people are apt to suppose. And, if much expenditure of bodily energy is involved in the day's work, mere knowledge is of still less importance when weighed against the probable cost of its acquirement. To do a fair day's work with his hands, a man needs, above all things, health, strength, and the patience and cheerfulness which, if they do not always accompany these blessings, can hardly in the nature of things exist without them; to which we must add honesty of purpose and a pride in doing what is done well.

A good handicraftsman can get on very well without genius, but he will fare badly without a reasonable share of that which is a more useful possession for workaday life, namely, mother-wit; and he will be all the better for a real knowledge, however limited, of the ordinary laws of nature, and especially of those which apply to his own business.

Instruction carried so far as to help the scholar to turn his store of mother-wit to account, to acquire a fair amount of sound elementary knowledge, and to use his hands and eyes; while leaving him fresh, vigorous, and with a sense of the dignity of his own calling, whatever it may be, if fairly and honestly pursued, cannot fail to be of invaluable service to all those who come under its influence.

But, on the other hand, if school instruction is carried so far as to encourage bookishness; if the ambition of the scholar is directed, not to the gaining of knowledge, but to the being able to pass examinations successfully; especially if encouragement is given to the mischievous delusion that brainwork is, in itself, and apart from its quality, a nobler or more respectable thing than handiwork—such education may be a deadly mischief to the workman, and lead to the rapid ruin of the industries it is intended to serve.

I know that I am expressing the opinion of some of the largest as well as the most enlightened employers of labour, when I say that there is a real danger that, from the extreme of no education, we may run to the other extreme of over-education of handicraftsmen. And I apprehend that what is true for the ordinary hand-worker is true for the foreman. Activity, probity, knowledge of men, ready mother-wit, supplemented by a good knowledge of the general principles involved in his business, are the making of a good foreman. If he possess these qualities, no amount of learning will fit him better for his position; while the course of life and the habit of mind required for the attainment of such learning may, in various direct and indirect ways, act as direct disqualifications for it.

Keeping in mind, then, that the two things to be avoided are, the delay of the entrance of boys into practical life, and the substitution of exhausted bookworms for shrewd, handy men, in our works and factories, let us consider what may be wisely and safely attempted in the way of improving the education of the handicraftsman.

First, I look to the elementary schools now happily established all over the country. I am not going to criticise or find fault with them; on the contrary, their establishment seems to me to be the most important and the most beneficial result of the corporate action of the people in our day. A great deal is said of British interests just now, but, depend upon it, that no Eastern difficulty needs our intervention as a nation so seriously, as the putting down both the Bashi-Bazouks of ignorance and the Cossacks of sectarianism at home. What has already been achieved in these directions is a great thing; you must have lived some time to know how great. An education, better in its processes, better in its substance, than that which was accessible to the great majority of well-to-do Britons a quarter of a century ago, is now obtainable by every child in the land. Let any man of my age go into an ordinary elementary school, and, unless he was unusually fortunate in his youth, he will tell you that the educational method, the intelligence, patience, and good temper on the teacher's part, which are now at the disposal of the veriest waifs and wastrels of society, are things of which he had no experience in those costly middle-class schools, which were so ingeniously contrived as to combine all the evils and shortcomings of the great public schools with none of their advantages. Many a man, whose so-called education cost a good deal of valuable money and occupied many a year of invaluable time, leaves the inspection of a well-ordered elementary school devoutly wishing that, in his young days, he had had the chance of being as well taught as these boys and girls are.

But while, in view of such an advance in general education, I willingly obey the natural impulse to be thankful, I am not willing altogether to rest. I want to see instruction in elementary science and in art more thoroughly incorporated in the educational system. At present, it is being administered by driblets, as if it were a potent medicine, "a few drops to be taken occasionally in a teaspoon." Every year I notice that that earnest and untiring friend of yours and of mine, Sir John Lubbock, stirs up the Government of the day in the House of Commons on this subject; and also that, every year, he, and the few members of the House of Commons, such as Mr. Playfair, who sympathise with him, are met with expressions of warm admiration for science in general, and reasons at large for doing nothing in particular. But now that Mr. Forster, to whom the education of the country owes so much, has announced his conversion to the right faith, I begin to hope that, sooner or later, things will mend.

I have given what I believe to be a good reason for the assumption, that the keeping at school of boys, who are to be handicraftsmen, beyond the age of thirteen or fourteen is neither practicable nor desirable; and, as it is quite certain, that, with justice to other and no less important branches of education, nothing more than the rudiments of science and art teaching can be introduced into elementary schools, we must seek elsewhere for a supplementary training in these subjects, and, if need be, in foreign languages, which may go on after the workman's life has begun.

The means of acquiring the scientific and artistic part of this training already exists in full working order, in the first place, in the classes of the Science and Art Department, which are, for the most part, held in the evening, so as to be accessible to all who choose to avail themselves of them after working hours. The great advantage of these classes is that they bring the means of instruction to the doors of the factories and workshops; that they are no artificial creations, but by their very existence prove the desire of the people for them; and finally, that they admit of indefinite development in proportion as they are wanted. I have often expressed the opinion, and I repeat it here, that, during the eighteen years they have been in existence, these classes have done incalculable good; and I can say, of my own knowledge, that the Department spares no pains and trouble in trying to increase their usefulness and ensure the soundness of their work.

No one knows better than my friend Colonel Donnelly, to whose clear views and great administrative abilities so much of the successful working of the science classes is due, that there is much to be done before the system can be said to be thoroughly satisfactory. The instruction given needs to be made more systematic and especially more practical; the teachers are of very unequal excellence, and not a few stand much in need of instruction themselves, not only in the subjects which they teach, but in the objects for which they teach. I daresay you have heard of that proceeding, reprobated by all true sportsmen, which is called "shooting for the pot." Well, there is such a thing as "teaching for the pot"—teaching, that is, not that your scholar may know, but that he may count for payment among those who pass the examination; and there are some teachers, happily not many, who have yet to learn that the examiners of the Department regard them as poachers of the worst description.

Without presuming in any way to speak in the name of the Department, I think I may say, as a matter which has come under my own observation, that it is doing its best to meet all these difficulties. It systematically promotes practical instruction in the classes; it affords facilities to teachers who desire to learn their business thoroughly; and it is always ready to aid in the suppression of pot-teaching.

All this is, as you may imagine, highly satisfactory to me. I see that spread of scientific education, about which I have so often permitted myself to worry the public, become, for all practical purposes, an accomplished fact. Grateful as I am for all that is now being done, in the same direction, in our higher schools and universities, I have ceased to have any anxiety about the wealthier classes. Scientific knowledge is spreading by what the alchemists called a "distillatio per ascensum;" and nothing now can prevent it from continuing to distil upwards and permeate English society, until, in the remote future, there shall be no member of the legislature who does not know as much of science as an elementary school-boy; and even the heads of houses in our venerable seats of learning shall acknowledge that natural science is not merely a sort of University back-door through which inferior men may get at their degrees. Perhaps this apocalyptic vision is a little wild; and I feel I ought to ask pardon for an outbreak of enthusiasm, which, I assure you, is not my commonest failing.

I have said that the Government is already doing a great deal in aid of that kind of technical education for handicraftsmen which, to my mind, is alone worth seeking. Perhaps it is doing as much as it ought to do, even in this direction. Certainly there is another kind of help of the most important character, for which we may look elsewhere than to the Government. The great mass of mankind have neither the liking, nor the aptitude, for either literary, or scientific, or artistic pursuits; nor, indeed, for excellence of any sort. Their ambition is to go through life with moderate exertion and a fair share of ease, doing common things in a common way. And a great blessing and comfort it is that the majority of men are of this mind; for the majority of things to be done are common things, and are quite well enough done when commonly done. The great end of life is not knowledge but action. What men need is, as much knowledge as they can assimilate and organise into a basis for action; give them more and it may become injurious. One knows people who are as heavy and stupid from undigested learning as others are from over-fulness of meat and drink. But a small percentage of the population is born with that most excellent quality, a desire for excellence, or with special aptitudes of some sort or another; Mr. Galton tells us that not more than one in four thousand may be expected to attain distinction, and not more than one in a million some share of that intensity of instinctive aptitude, that burning thirst for excellence, which is called genius.

Now, the most important object of all educational schemes is to catch these exceptional people, and turn them to account for the good of society. No man can say where they will crop up; like their opposites, the fools and knaves, they appear sometimes in the palace, and sometimes in the hovel; but the great thing to be aimed at, I was almost going to say the most important end of all social arrangements, is to keep these glorious sports of Nature from being either corrupted by luxury or starved by poverty, and to put them into the position in which they can do the work for which they are specially fitted.

Thus, if a lad in an elementary school showed signs of special capacity, I would try to provide him with the means of continuing his education after his daily working life had begun; if, in the evening classes, he developed special capabilities in the direction of science or of drawing, I would try to secure him an apprenticeship to some trade in which those powers would have applicability. Or, if he chose to become a teacher, he should have the chance of so doing. Finally, to the lad of genius, the one in a million, I would make accessible the highest and most complete training the country could afford. Whatever that might cost, depend upon it the investment would be a good one. I weigh my words when I say that if the nation could purchase a potential Watt, or Davy, or Faraday, at the cost of a hundred thousand pounds down, he would be dirt-cheap at the money. It is a mere commonplace and everyday piece of knowledge, that what these three men did has produced untold millions of wealth, in the narrowest economical sense of the word.

Therefore, as the sum and crown of what is to be done for technical education, I look to the provision of a machinery for winnowing out the capacities and giving them scope. When I was a member of the London School Board, I said, in the course of a speech, that our business was to provide a ladder, reaching from the gutter to the university, along which every child in the three kingdoms should have the chance of climbing as far as he was fit to go. This phrase was so much bandied about at the time, that, to say truth, I am rather tired of it; but I know of no other which so fully expresses my belief, not only about education in general, but about technical education in particular.

The essential foundation of all the organisation needed for the promotion of education among handicraftsmen will, I believe, exist in this country, when every working lad can feel that society has done as much as lies in its power to remove all needless and artificial obstacles from his path; that there is no barrier, except such as exists in the nature of things, between himself and whatever place in the social organisation he is fitted to fill; and, more than this, that, if he has capacity and industry, a hand is held out to help him along any path which is wisely and honestly chosen.

I have endeavoured to point out to you that a great deal of such an organisation already exists; and I am glad to be able to add that there is a good prospect that what is wanting will, before long, be supplemented.

Those powerful and wealthy societies, the livery companies of the City of London, remembering that they are the heirs and representatives of the trade guilds of the Middle Ages, are interesting themselves in the question. So far back as 1872 the Society of Arts organised a system of instruction in the technology of arts and manufactures, for persons actually employed in factories and workshops, who desired to extend and improve their knowledge of the theory and practice of their particular avocations;[1] and a considerable subsidy, in aid of the efforts of the Society, was liberally granted by the Clothworkers' Company. We have here the hopeful commencement of a rational organisation for the promotion of excellence among handicraftsmen. Quite recently, other of the livery companies have determined upon giving their powerful, and, indeed, almost boundless, aid to the improvement of the teaching of handicrafts. They have already gone so far as to appoint a committee to act for them; and I betray no confidence in adding that, some time since, the committee sought the advice and assistance of several persons, myself among the number.

Of course I cannot tell you what may be the result of the deliberations of the committee; but we may all fairly hope that, before long, steps which will have a weighty and a lasting influence on the growth and

[1] See the "Programme" for 1878, issued by the Society of Arts, p. 14.

spread of sound and thorough teaching among the handicraftsmen[1] of this country will be taken by the livery companies of London.

[This hope has been fully justified by the establishment of the Cowper Street Schools, and that of the Central Institution of the City and Guilds of London Institute. September 1881.]

---

[1] It is perhaps advisable to remark that the important question of the professional education of managers of industrial works is not touched in the foregoing remarks.

IV.

ON ELEMENTARY INSTRUCTION IN PHYSIOLOGY.

The chief ground upon which I venture to recommend that the teaching of elementary physiology should form an essential part of any organised course of instruction in matters pertaining to domestic economy, is, that a knowledge of even the elements of this subject supplies those conceptions of the constitution and mode of action of the living body, and of the nature of health and disease, which prepare the mind to receive instruction from sanitary science.

It is, I think, eminently desirable that the hygienist and the physician should find something in the public mind to which they can appeal; some little stock of universally acknowledged truths, which may serve as a foundation for their warnings, and predispose towards an intelligent obedience to their recommendations.

Listening to ordinary talk about health, disease, and death, one is often led to entertain a doubt whether the speakers believe that the course of natural causation runs as smoothly in the human body as elsewhere. Indications are too often obvious of a strong, though perhaps an unavowed and half unconscious, under-current of opinion that the phenomena of life are not only widely different, in their superficial characters and in their practical importance, from other natural events, but that they do not follow in that definite order which characterises the succession of all other occurrences, and the statement of which we call a law of nature.

Hence, I think, arises the want of heartiness of belief in the value of knowledge respecting the laws of health and disease, and of the foresight and care to which knowledge is the essential preliminary, which is so often noticeable; and a corresponding laxity and carelessness in practice, the results of which are too frequently lamentable.

It is said that among the many religious sects of Russia, there is one which holds that all disease is brought about by the direct and special interference of the Deity, and which, therefore, looks with repugnance upon both preventive and curative measures as alike blasphemous interferences with the will of God. Among ourselves, the "Peculiar People" are, I believe, the only persons who hold the like doctrine in its integrity, and carry it out with logical rigour. But many of us are old enough to recollect that the administration of chloroform in assuagement of the pangs of childbirth was, at its introduction, strenuously resisted upon similar grounds.

I am not sure that the feeling, of which the doctrine to which I have referred is the full expression, does not lie at the bottom of the minds of a great many people who yet would vigorously object to give a verbal assent to the doctrine itself. However this may be, the main point is that sufficient knowledge has now been acquired of vital phenomena, to justify the assertion, that the notion, that there is anything exceptional about these phenomena, receives not a particle of support from any known fact. On the contrary, there is a vast and an increasing mass of evidence that birth and death, health and disease, are as much parts of the ordinary stream of events as the rising and setting of the sun, or the changes of the moon; and that the living body is a mechanism, the proper working of which we term health; its disturbance, disease; its stoppage, death. The activity of this mechanism is dependent upon many and complicated conditions, some of which are hopelessly beyond our control, while others are readily accessible, and are capable of being indefinitely modified by our own actions. The business of the hygienist and of the physician is to know the range of these modifiable conditions, and how to influence them towards the maintenance of health and the prolongation of life; the business of the general public is to give an intelligent assent, and a ready obedience based upon that assent, to the rules laid down for their guidance by such experts. But an intelligent assent is an assent based upon knowledge, and the knowledge which is here in question means an acquaintance with the elements of physiology.

It is not difficult to acquire such knowledge. What is true, to a certain extent, of all the physical sciences, is eminently characteristic of physiology—the difficulty of the subject begins beyond the stage of elementary knowledge, and increases with every stage of progress. While the most highly trained and the best furnished intellect may find all its resources insufficient, when it strives to reach the heights and penetrate into the depths of the problems of physiology, the elementary and fundamental truths can be made clear to a child.

No one can have any difficulty in comprehending the mechanism of circulation or respiration; or the general mode of operation of the organ of vision; though the unravelling of all the minutiæ of these processes, may, for the present, baffle the conjoined attacks of the most accomplished physicists, chemists, and mathematicians. To know the anatomy of the human body, with even an approximation to thoroughness, is the work of a life; but as much as is needed for a sound comprehension of elementary physiological truths, may be learned in a week.

A knowledge of the elements of physiology is not only easy of acquirement, but it may be made a real and practical acquaintance with the facts, as far as it goes. The subject of study is always at hand, in oneself. The principal constituents of the skeleton, and the changes of form of contracting muscles, may be felt through one's own skin. The beating of one's heart, and its connection with the pulse, may be noted; the influence of the valves of one's own veins may be shown; the movements of respiration may be observed; while the wonderful phenomena of sensation afford an endless field for curious and interesting self-study. The prick of a needle will yield, in a drop of one's own blood, material for microscopic observation of phenomena which lie at the foundation of all biological conceptions; and a cold, with its concomitant coughing and sneezing, may prove the sweet uses of adversity by helping one to a clear conception of what is meant by "reflex action."

Of course there is a limit to this physiological self-examination. But there is so close a solidarity between ourselves and our poor relations of the animal world, that our inaccessible inward parts may be supplemented by theirs. A comparative anatomist knows that a sheep's heart and lungs, or eye, must not be confounded with those of a man; but, so far as the comprehension of the elementary facts of the physiology of circulation, of respiration, and of vision goes, the one furnishes the needful anatomical data as well as the other.

Thus, it is quite possible to give instruction in elementary physiology in such a manner as, not only to confer knowledge, which, for the reason I have mentioned, is useful in itself; but to serve the purposes of a training in accurate observation, and in the methods of reasoning of physical science. But that is an advantage which I mention only incidentally, as the present Conference does not deal with education in the ordinary sense of the word.

It will not be suspected that I wish to make physiologists of all the world. It would be as reasonable to accuse an advocate of the "three R's" of a desire to make an orator, an author, and a mathematician of everybody. A stumbling reader, a pot-hook writer, and an arithmetician who has not got beyond the rule of three, is not a person of brilliant acquirements; but the difference between such a member of society and one who can neither read, write, nor cipher is almost inexpressible; and no one now-a-days doubts the value of instruction, even if it goes no farther.

The saying that a little knowledge is a dangerous thing is, to my mind, a very dangerous adage. If knowledge is real and genuine, I do not believe that it is other than a very valuable possession, however infinitesimal its quantity may be. Indeed, if a little knowledge is dangerous, where is the man who has so much as to be out of danger?

If William Harvey's life-long labours had revealed to him a tenth part of that which may be made sound and real knowledge to our boys and girls, he would not only have been what he was, the greatest physiologist of his age, but he would have loomed upon the seventeenth century as a sort of intellectual portent. Our "little knowledge" would have been to him a great, astounding, unlooked-for vision of scientific truth.

I really see no harm which can come of giving our children a little knowledge of physiology. But then, as I have said, the instruction must be real, based upon observation, eked out by good explanatory diagrams and models, and conveyed by a teacher whose own knowledge has been acquired by a study of the facts; and not the mere catechismal parrot-work which too often usurps the place of elementary teaching.

It is, I hope, unnecessary for me to give a formal contradiction to the silly fiction, which is assiduously circulated by fanatics who not only ought to know, but do know, that their assertions are untrue, that I have advocated the introduction of that experimental discipline which is absolutely indispensable to the professed physiologist, into elementary teaching.

But while I should object to any experimentation which can justly be called painful, for the purpose of elementary instruction; and, while, as a member of a late Royal Commission, I gladly did my best to prevent the infliction of needless pain, for any purpose; I think it is my duty to take this opportunity of expressing my regret at a condition of the law which permits a boy to troll for pike, or set lines with live frog bait, for idle amusement; and, at the same time, lays the teacher of that boy open to the penalty of fine and imprisonment, if he uses the same animal for the purpose of exhibiting one of the most beautiful and instructive of physiological spectacles, the circulation in the web of the foot. No one could undertake to affirm that a frog is not inconvenienced by being wrapped up in a wet rag, and having his toes tied out; and it cannot be denied that inconvenience is a sort of pain. But you must not inflict the least pain on a vertebrated animal for scientific purposes (though you may do a good deal in that way for gain or for sport) without due licence of the Secretary of State for the

Home Department, granted under the authority of the Vivisection Act.

So it comes about, that, in this present year of grace 1877, two persons may be charged with cruelty to animals. One has impaled a frog, and suffered the creature to writhe about in that condition for hours; the other has pained the animal no more than one of us would be pained by tying strings round his fingers, and keeping him in the position of a hydropathic patient. The first offender says, "I did it because I find fishing very amusing," and the magistrate bids him depart in peace; nay, probably wishes him good sport. The second pleads, "I wanted to impress a scientific truth, with a distinctness attainable in no other way, on the minds of my scholars," and the magistrate fines him five pounds.

I cannot but think that this is an anomalous and not wholly creditable state of things.

V.

JOSEPH PRIESTLEY.

If the man to perpetuate whose memory we have this day raised a statue had been asked on what part of his busy life's work he set the highest value, he would undoubtedly have pointed to his voluminous contributions to theology. In season and out of season, he was the steadfast champion of that hypothesis respecting the Divine nature which is termed Unitarianism by its friends and Socinianism by its foes. Regardless of odds, he was ready to do battle with all comers in that cause; and if no adversaries entered the lists, he would sally forth to seek them.

To this, his highest ideal of duty, Joseph Priestley sacrificed the vulgar prizes of life, which, assuredly, were within easy reach of a man of his singular energy and varied abilities. For this object, he put aside, as of secondary importance, those scientific investigations which he loved so well, and in which he showed himself so competent to enlarge the boundaries of natural knowledge and to win fame. In this cause, he not only cheerfully suffered obloquy from the bigoted and the unthinking, and came within sight of martyrdom; but bore with that which is much harder to be borne than all these, the unfeigned astonishment and hardly disguised contempt of a brilliant society, composed of men whose sympathy and esteem must have been most dear to him, and to whom it was simply incomprehensible that a philosopher should seriously occupy himself with any form of Christianity.

It appears to me that the man who, setting before himself such an ideal of life, acted up to it consistently, is worthy of the deepest respect, whatever opinion may be entertained as to the real value of the tenets which he so zealously propagated and defended.

But I am sure that I speak not only for myself, but for all this assemblage, when I say that our purpose to-day is to do honour, not to Priestley, the Unitarian divine, but to Priestley, the fearless defender of rational freedom in thought and in action: to Priestley, the philosophic thinker; to that Priestley who held a foremost place among "the swift runners who hand over the lamp of life,"[1] and transmit from one generation to another the fire kindled, in the childhood of the world, at the Promethean altar of Science.

The main incidents of Priestley's life are so well known that I need dwell upon them at no great length.

Born in 1733, at Fieldhead, near Leeds, and brought up among Calvinists of the straitest orthodoxy, the boy's striking natural ability led to his

[1] "Quasi cursores, vitai lampada tradunt." — LUCR. *De Rerum Nat.* ii. 78.

being devoted to the profession of a minister of religion; and, in 1752, he was sent to the Dissenting Academy at Daventry—an institution which authority left undisturbed, though its existence contravened the law. The teachers under whose instruction and influence the young man came at Daventry, carried out to the letter the injunction to "try all things: hold fast that which is good," and encouraged the discussion of every imaginable proposition with complete freedom, the leading professors taking opposite sides; a discipline which, admirable as it may be from a purely scientific point of view, would seem to be calculated to make acute, rather than sound, divines. Priestley tells us, in his "Autobiography," that he generally found himself on the unorthodox side: and, as he grew older, and his faculties attained their maturity, this native tendency towards heterodoxy grew with his growth and strengthened with his strength. He passed from Calvinism to Arianism; and finally, in middle life, landed in that very broad form of Unitarianism, by which his craving after a credible and consistent theory of things was satisfied.

On leaving Daventry, Priestley became minister of a congregation, first at Needham Market, and secondly at Nantwich; but whether on account of his heterodox opinions, or of the stuttering which impeded his expression of them in the pulpit, little success attended his efforts in this capacity. In 1761, a career much more suited to his abilities became open to him. He was appointed "tutor in the languages" in the Dissenting Academy at Warrington, in which capacity, besides giving three courses of lectures, he taught Latin, Greek, French, and Italian, and read lectures on the Theory of Language and Universal Grammar, on Oratory, Philosophical Criticism, and Civil law. And it is interesting to observe that, as a teacher, he encouraged and cherished in those whom he instructed, the freedom which he had enjoyed, in his own student days, at Daventry. One of his pupils tells us that,

> "At the conclusion of his lecture, he always encouraged his students to express their sentiments relative to the subject of it, and to urge any objections to what he had delivered, without reserve. It pleased him when any one commenced such a conversation. In order to excite the freest discussion, he occasionally invited the students to drink tea with him, in order to canvass the subjects of his lectures. I do not recollect that he ever showed the least displeasure at the strongest objections that were made to what he delivered, but I distinctly remember the smile of approbation with which he usually received them: nor did he fail to point out, in a very encouraging manner, the ingenuity or force of any remarks that were made, when they merited these characters. His object, as well as Dr. Aikin's, was to engage the students to examine and decide for themselves, uninfluenced by the sentiments of any other persons."[1]

It would be difficult to give a better description of a model teacher than that conveyed in these words.

From his earliest days, Priestley had shown a strong bent towards the study of nature; and his brother Timothy tells us that the boy put spiders into bottles, to see how long they would live in the same air—a curious anticipation of the investigations

[1] "Life and Correspondence of Dr. Priestley," by J. T. Rutt. Vol. i. p. 50.

of his later years. At Nantwich, where he set up a school, Priestley informs us that he bought an air pump, an electrical machine, and other instruments, in the use of which he instructed his scholars. But he does not seem to have devoted himself seriously to physical science until 1766, when he had the great good fortune to meet Benjamin Franklin, whose friendship he ever afterwards enjoyed. Encouraged by Franklin, he wrote a "History of Electricity," which was published in 1767, and appears to have met with considerable success.

In the same year, Priestley left Warrington to become the minister of a congregation at Leeds; and, here, happening to live next door to a public brewery, as he says,

"I, at first, amused myself with making experiments on the fixed air which I found ready-made in the process of fermentation. When I removed from that house I was under the necessity of making fixed air for myself; and one experiment leading to another, as I have distinctly and faithfully noted in my various publications on the subject, I by degrees contrived a convenient apparatus for the purpose, but of the cheapest kind.

"When I began these experiments I knew very little of *chemistry*, and had, in a manner, no idea on the subject before I attended a course of chemical lectures, delivered in the Academy at Warrington, by Dr. Turner of Liverpool. But I have often thought that, upon the whole, this circumstance was no disadvantage to me; as, in this situation, I was led to devise an apparatus and processes of my own, adapted to my peculiar views; whereas, if I had been previously accustomed to the usual chemical processes, I should not have so easily thought of any other, and without new modes of operation, I should hardly have discovered anything materially new."[1]

[1] "Autobiography," §§ 100, 101.

The first outcome of Priestley's chemical work, published in 1772, was of a very practical character. He discovered the way of impregnating water with an excess of "fixed air," or carbonic acid, and thereby producing what we now know as "soda water"—a service to naturally, and still more to artificially, thirsty souls, which those whose parched throats and hot heads are cooled by morning draughts of that beverage, cannot too gratefully acknowledge. In the same year, Priestley communicated the extensive series of observations which his industry and ingenuity had accumulated, in the course of four years, to the Royal Society, under the title of "Observations on Different Kinds of Air"—a memoir which was justly regarded of so much merit and importance, that the Society at once conferred upon the author the highest distinction in their power, by awarding him the Copley Medal.

In 1771 a proposal was made to Priestley to accompany Captain Cook in his second voyage to the South Seas. He accepted it, and his congregation agreed to pay an assistant to supply his place during his absence. But the appointment lay in the hands of the Board of Longitude, of which certain clergymen were members; and whether these worthy ecclesiastics feared that Priestley's presence among the ship's company might expose his Majesty's Sloop *Resolution* to the fate which aforetime befell a certain ship that went from Joppa to Tarshish; or whether they were alarmed lest a Socinian should undermine that piety which, in the days of Commodore Trunnion, so strikingly characterised sailors, does not appear; but, at any rate, they objected to Priestley "on account of his religious principles," and appointed the two Forsters, whose "religious principles," if they had been known to these well-meaning but not far-sighted persons, would probably have surprised them.

In 1772 another proposal was made to Priestley. Lord Shelburne, desiring a "literary companion," had been brought into communication with Priestley by the good offices of a friend of both, Dr. Price; and offered him the nominal post of librarian, with a good house and appointments, and an annuity in case of the termination of the engagement. Priestley accepted the offer, and remained with Lord Shelburne for seven years, sometimes residing at Calne, sometimes travelling abroad with the Earl.

Why the connection terminated has never been exactly known; but it is certain that Lord Shelburne behaved with the utmost consideration and kindness towards Priestley; that he fulfilled his engagements to the letter; and that, at a later period, he expressed a desire that Priestley should return to his old footing in his house. Probably enough, the politician, aspiring to the highest offices in the state, may have found the position of the protector of a man who was being denounced all over the country as an infidel and an atheist somewhat embarrassing. In fact, a passage in Priestley's "Autobiography" on the occasion of the publication of his "Disquisitions relating to Matter and Spirit," which took place in 1777, indicates pretty clearly the state of the case:—

"(126) It being probable that this publication would be unpopular, and might be the means of bringing odium on my patron, several attempts were made by his friends, though none by himself, to dissuade me from persisting in it. But being, as I thought, engaged in the cause of important truth, I proceeded without regard to any consequences, assuring them that this publication should not be injurious to his lordship."

It is not unreasonable to suppose that his lordship, as a keen, practical man of the world, did not derive much satisfaction from this assurance. The "evident marks of dissatisfaction" which Priestley says he first perceived in his patron in 1778, may well have arisen from the peer's not unnatural uneasiness as to what his domesticated, but not tamed, philosopher might write next, and what storm might thereby be brought down on his own head; and it speaks very highly for Lord Shelburne's delicacy that, in the midst of such perplexities, he made not the least attempt to interfere with Priestley's freedom of action. In 1780, however, he intimated to Dr. Price that he should be glad to establish Priestley on his Irish estates: the suggestion was interpreted, as Lord Shelburne probably intended it should be, and Priestley left him, the annuity of £150 a year, which had been promised in view of such a contingency, being punctually paid.

After leaving Calne, Priestley spent some little time in London, and then, having settled in Birmingham at the desire of his brother-in-law, he was soon invited to become the minister of a large congregation. This settlement Priestley considered, at the time, to be "the happiest event of his life." And well he might think so; for it gave him competence and leisure; placed him within reach of the best makers of apparatus of the day; made him a member of that remarkable "Lunar Society," at whose meetings he could exchange thoughts with such men as Watt, Wedgewood, Darwin, and Boulton; and threw open to him the pleasant house of the Galtons of Barr, where these men, and others of less note, formed a society of exceptional charm and intelligence.[1]

But these halcyon days were ended by a bitter storm. The French Revolution broke out. An electric shock ran through the nations; whatever there was of corrupt and retrograde, and, at the same time, a great deal of what there was of best and noblest, in European society shuddered at the outburst of long-pent-up social fires. Men's feelings were excited in a way that we, in this generation, can hardly comprehend. Party wrath and virulence were expressed in a manner unparalleled, and it is to be

[1] See "The Life of Mary Anne Schimmelpenninck." Mrs. Schimmelpenninck (*née* Galton) remembered Priestley very well, and her description of him is worth quotation :—"A man of admirable simplicity, gentleness and kindness of heart, united with great acuteness of intellect. I can never forget the impression produced on me by the serene expression of his countenance. He, indeed, seemed present with God by recollection, and with man by cheerfulness. I remember that, in the assembly of these distinguished men, amongst whom Mr. Boulton, by his noble manner, his fine countenance (which much resembled that of Louis XIV.), and princely munificence, stood pre-eminently as the great Mecænas; even as a child, I used to feel, when Dr. Priestley entered after him, that the glory of the one was terrestrial, that of the other celestial; and utterly far as I am removed from a belief in the sufficiency of Dr. Priestley's theological creed, I cannot but here record this evidence of the eternal power of any portion of the truth held in its vitality."

hoped impossible, in our times; and Priestley and his friends were held up to public scorn, even in Parliament, as fomenters of sedition. A "Church-and-King" cry was raised against the Liberal Dissenters; and, in Birmingham, it was intensified and specially directed towards Priestley by a local controversy, in which he had engaged with his usual vigour. In 1791, the celebration of the second anniversary of the taking of the Bastille by a public dinner, with which Priestley had nothing whatever to do, gave the signal to the loyal and pious mob, who, unchecked, and indeed to some extent encouraged, by those who were responsible for order, had the town at their mercy for three days. The chapels and houses of the leading Dissenters were wrecked, and Priestley and his family had to fly for their lives, leaving library, apparatus, papers, and all their possessions, a prey to the flames.

Priestley never returned to Birmingham. He bore the outrages and losses inflicted upon him with extreme patience and sweetness,[1] and betook himself to London. But even his scientific colleagues gave him a cold shoulder; and though he was elected minister of a congregation at Hackney, he felt his position to be insecure, and finally determined on emigrating to the United States. He landed in America in 1794; lived quietly with his sons at Northumberland, in

[1] Even Mrs. Priestley, who might be forgiven for regarding the destroyers of her household gods with some asperity, contents herself, in writing to Mrs. Barbauld, with the sarcasm that the Birmingham people "will scarcely find so many respectable characters, a second time, to make a bonfire of."

Pennsylvania, where his posterity still flourish; and, clear-headed and busy to the last, died on the 6th of February 1804.

Such were the conditions under which Joseph Priestley did the work which lay before him, and then, as the Norse Sagas say, went out of the story. The work itself was of the most varied kind. No human interest was without its attraction for Priestley, and few men have ever had so many irons in the fire at once; but, though he may have burned his fingers a little, very few who have tried that operation have burned their fingers so little. He made admirable discoveries in science; his philosophical treatises are still well worth reading; his political works are full of insight and replete with the spirit of freedom; and while all these sparks flew off from his anvil, the controversial hammer rained a hail of blows on orthodox priest and bishop. While thus engaged, the kindly, cheerful doctor felt no more wrath or uncharitableness towards his opponents than a smith does towards his iron. But if the iron could only speak!—and the priests and bishops took the point of view of the iron.

No doubt what Priestley's friends repeatedly urged upon him—that he would have escaped the heavier trials of his life and done more for the advancement of knowledge, if he had confined himself to his scientific pursuits and let his fellow-men go their way —was true. But it seems to have been Priestley's feeling that he was a man and a citizen before he was a philosopher, and that the duties of the two former positions are at least as imperative as those of the latter. Moreover, there are men (and I think Priestley was one of them) to whom the satisfaction of throwing down a triumphant fallacy is as great as that which attends the discovery of a new truth; who feel better satisfied with the government of the world, when they have been helping Providence by knocking an imposture on the head; and who care even more for freedom of thought than for mere advance of knowledge. These men are the Carnots who organise victory for truth, and they are, at least, as important as the generals who visibly fight her battles in the field.

Priestley's reputation as a man of science rests upon his numerous and important contributions to the chemistry of gaseous bodies; and to form a just estimate of the value of his work—of the extent to which it advanced the knowledge of fact and the development of sound theoretical views—we must reflect what chemistry was in the first half of the eighteenth century.

The vast science which now passes under that name had no existence. Air, water, and fire were still counted among the elemental bodies; and though Van Helmont, a century before, had distinguished different kinds of air as *gas ventosum* and *gas sylvestre*, and Boyle and Hales had experimentally defined the physical properties of air, and discriminated some of the various kinds of aëriform bodies, no one suspected the existence of the numerous totally distinct gaseous elements which are now known, or dreamed that the air we breathe and the water we drink are compounds of gaseous elements.

But, in 1754, a young Scotch physician, Dr. Black, made the first clearing in this tangled backwood of knowledge. And it gives one a wonderful impression of the juvenility of scientific chemistry to think that Lord Brougham, whom so many of us recollect, attended Black's lectures when he was a student in Edinburgh. Black's researches gave the world the novel and startling conception of a gas that was a permanently elastic fluid like air, but that differed from common air in being much heavier, very poisonous, and in having the properties of an acid, capable of neutralising the strongest alkalies; and it took the world some time to become accustomed to the notion.

A dozen years later, one of the most sagacious and accurate investigators who has adorned this, or any other, country, Henry Cavendish, published a memoir in the "Philosophical Transactions," in which he deals not only with the "fixed air" (now called carbonic acid or carbonic anhydride) of Black, but with "inflammable air," or what we now term hydrogen.

By the rigorous application of weight and measure to all his processes, Cavendish implied the belief subsequently formulated by Lavoisier, that, in chemical processes, matter is neither created nor destroyed, and indicated the path along which all future explorers must travel. Nor did he himself halt until this path led him, in 1784, to the brilliant and fundamental discovery that water is composed of two gases united in fixed and constant proportions.

It is a trying ordeal for any man to be compared with Black and Cavendish, and Priestley cannot be said to stand on their level. Nevertheless, his achievements are not only great in themselves, but truly wonderful, if we consider the disadvantages under which he laboured. Without the careful scientific training of Black, without the leisure and appliances secured by the wealth of Cavendish, he scaled the walls of science as so many Englishmen have done before and since his day; and trusting to mother wit to supply the place of training, and to ingenuity to create apparatus out of washing tubs, he discovered more new gases than all his predecessors put together had done. He laid the foundations of gas analysis; he discovered the complementary actions of animal and vegetable life upon the constituents of the atmosphere; and, finally, he crowned his work, this day one hundred years ago, by the discovery of that "pure dephlogisticated air" to which the French chemists subsequently gave the name of oxygen. Its importance, as the constituent of the atmosphere which disappears in the processes of respiration and combustion, and is restored by green plants growing in sunshine, was proved somewhat later. For these brilliant discoveries, the Royal Society elected Priestley a fellow and gave him their medal, while the Academies of Paris and St. Petersburg conferred their membership upon him. Edinburgh had made him an honorary doctor of laws at an early period of his career; but, I need hardly add, that a man of Priestley's opinions received no recognition from the universities of his own country.

That Priestley's contributions to the knowledge of chemical fact were of the greatest importance, and that they richly deserve all the praise that has been awarded to them, is unquestionable; but it must, at the same time, be admitted that he had no comprehension of the deeper significance of his work; and, so far from contributing anything to the theory of the facts which he discovered, or assisting in their rational explanation, his influence to the end of his life was warmly exerted in favour of error. From first to last, he was a stiff adherent of the phlogiston doctrine which was prevalent when his studies commenced; and, by a curious irony of fate, the man who by the discovery of what he called "dephlogisticated air" furnished the essential datum for the true theory of combustion, of respiration, and of the composition of water, to the end of his days fought against the inevitable corollaries from his own labours. His last scientific work, published in 1800, bears the title, "The Doctrine of Phlogiston established, and that of the Composition of Water refuted."

When Priestley commenced his studies, the current belief was, that atmospheric air, freed from accidental impurities, is a simple elementary substance, indestructible and unalterable, as water was supposed to be. When a combustible burned, or when an animal breathed in air, it was supposed that a substance, "phlogiston," the matter of heat and light, passed from the burning or breathing body into it, and destroyed its powers of supporting life and combustion. Thus, air contained in a vessel in which a lighted candle had gone out, or a living animal had breathed until it could breathe no longer, was called "phlogisticated." The same result was supposed to be brought about by the addition of what Priestley called "nitrous gas" to common air.

In the course of his researches, Priestley found that the quantity of common air which can thus become "phlogisticated," amounts to about one-fifth the volume of the whole quantity submitted to experiment. Hence it appeared that common air consists, to the extent of four-fifths of its volume, of air which is already "phlogisticated;" while the other fifth is free from phlogiston, or "dephlogisticated." On the other hand, Priestley found that air "phlogisticated" by combustion or respiration could be "dephlogisticated," or have the properties of pure common air restored to it, by the action of green plants in sunshine. The question, therefore, would naturally arise —as common air can be wholly phlogisticated by combustion, and converted into a substance which will no longer support combustion, is it possible to get air that shall be less phlogisticated than common air, and consequently support combustion better than common air does?

Now, Priestley says that, in 1774, the possibility of obtaining air less phlogisticated than common air had not occurred to him.[1] But in pursuing his experiments on the evolution of air from various bodies by means of heat, it happened that, on the 1st of August 1774, he threw the heat of the sun, by means of a large burning glass which he had recently obtained, upon a substance which was then called *mercurius calcinatus per se*, and which is commonly known as red precipitate.

"I presently found that, by means of this lens, air was expelled from it very readily. Having got about three or four times as much as the bulk of my materials, I admitted water to it, and found that it was not imbibed by it. But what surprised me more than I can well express, was that a candle burned in this air with a remarkably vigorous flame, very much like that enlarged flame with which a candle burns in nitrous air, exposed to iron or lime of sulphur; but as I had got nothing like this remarkable appearance from any kind of air besides this particular modification of nitrous air, and I knew no nitrous acid was used in the preparation of *mercurius calcinatus*, I was utterly at a loss how to account for it.

"In this case also, though I did not give sufficient attention to the circumstance at that time, the flame of the candle, besides being larger, burned with more splendour and heat than in that species of nitrous air; and a piece of red-hot wood sparkled in it, exactly like paper dipped in a solution of nitre, and it consumed very fast—an experiment which I had never thought of trying with nitrous air."[2]

Priestley obtained the same sort of air from red lead, but, as he says himself, he remained in ignorance of the properties of this new kind of air for seven months, or until March 1775, when he found

[1] "Experiments and Observations on Different Kinds of Air," vol. ii. p. 31.

[2] *Ibid.* pp. 34, 35.

that the new air behaved with "nitrous gas" in the same way as the dephlogisticated part of common air does;[1] but that, instead of being diminished to four-fifths, it almost completely vanished, and, therefore, showed itself to be "between five and six times as good as the best common air I have ever met with."[2] As this new air thus appeared to be completely free from phlogiston, Priestley called it "dephlogisticated air."

What was the nature of this air? Priestley found that the same kind of air was to be obtained by moistening with the spirit of nitre (which he terms nitrous acid) any kind of earth that is free from phlogiston, and applying heat; and consequently he says: "There remained no doubt on my mind but that the atmospherical air, or the thing that we breathe, consists of the nitrous acid and earth, with so much phlogiston as is necessary to its elasticity, and likewise so much more as is required to bring it from its state of perfect purity to the mean condition in which we find it."[3]

Priestley's view, in fact, is that atmospheric air is a kind of saltpetre, in which the potash is replaced by some unknown earth. And in speculating on the manner in which saltpetre is formed, he enunciates the hypothesis, "that nitre is formed by a real *decomposition of the air itself*, the *bases* that are presented to it having, in such circumstances, a nearer

[1] "Experiments and Observations on Different Kinds of Air," vol. ii. p. 40. [2] *Ibid.* p. 48. [3] *Ibid.* p. 55.

affinity with the spirit of nitre than that kind of earth with which it is united in the atmosphere."[1]

It would have been hard for the most ingenious person to have wandered farther from the truth than Priestley does in this hypothesis; and, though Lavoisier undoubtedly treated Priestley very ill, and pretended to have discovered dephlogisticated air, or oxygen, as he called it, independently, we can almost forgive him, when we reflect how different were the ideas which the great French chemist attached to the body which Priestley discovered.

They are like two navigators of whom the first sees a new country, but takes clouds for mountains and mirage for lowlands; while the second determines its length and breadth, and lays down on a chart its exact place, so that, thenceforth, it serves as a guide to his successors, and becomes a secure outpost whence new explorations may be pushed.

Nevertheless, as Priestley himself somewhere remarks, the first object of physical science is to ascertain facts, and the service which he rendered to chemistry by the definite establishment of a large number of new and fundamentally important facts, is such as to entitle him to a very high place among the fathers of chemical science.

It is difficult to say whether Priestley's philosophical, political, or theological views were most responsible for the bitter hatred which was borne to

[1] "Experiments and Observations on Different Kinds of Air," vol. ii. p. 60. The italics are Priestley's own.

him by a large body of his countrymen,[1] and which found its expression in the malignant insinuations in which Burke, to his everlasting shame, indulged in the House of Commons.

Without containing much that will be new to the readers of Hobbes, Spinoza, Collins, Hume, and Hartley, and, indeed, while making no pretensions to originality, Priestley's "Disquisitions relating to Matter and Spirit," and his "Doctrine of Philosophical Necessity illustrated," are among the most powerful, clear, and unflinching expositions of materialism and necessarianism which exist in the English language, and are still well worth reading.

Priestley denied the freedom of the will in the sense of its self-determination; he denied the existence of a soul distinct from the body; and as a natural consequence, he denied the natural immortality of man.

In relation to these matters English opinion, a century ago, was very much what it is now.

A man may be a necessarian without incurring

[1] "In all the newspapers and most of the periodical publications I was represented as an unbeliever in Revelation, and no better than an atheist."—"Autobiography," Rutt. vol. i. p. 124. "On the walls of houses, etc., and especially where I usually went, were to be seen, in large characters, 'MADAN FOR EVER; DAMN PRIESTLEY; NO PRESBYTERIANISM; DAMN THE PRESBYTERIANS,' etc. etc.; and, at one time, I was followed by a number of boys, who left their play, repeating what they had seen on the walls, and shouting out, '*Damn Priestley; damn him, damn him, for ever, for ever,*' etc. etc. This was no doubt a lesson which they had been taught by their parents, and what they, I fear, had learned from their superiors."—"Appeal to the Public on the Subject of the Riots at Birmingham."

graver reproach than that implied in being called a gloomy fanatic, necessarianism, though very shocking, having a note of Calvinistic orthodoxy; but, if a man is a materialist; or, if good authorities say he is and must be so, in spite of his assertion to the contrary; or, if he acknowledge himself unable to see good reasons for believing in the natural immortality of man, respectable folks look upon him as an unsafe neighbour of a cash-box, as an actual or potential sensualist, the more virtuous in outward seeming, the more certainly loaded with secret "grave personal sins."

Nevertheless, it is as certain as anything can be, that Joseph Priestley was no gloomy fanatic, but as cheerful and kindly a soul as ever breathed, the idol of children; a man who was hated only by those who did not know him, and who charmed away the bitterest prejudices in personal intercourse; a man who never lost a friend, and the best testimony to whose worth is the generous and tender warmth with which his many friends vied with one another in rendering him substantial help, in all the crises of his career.

The unspotted purity of Priestley's life, the strictness of his performance of every duty, his transparent sincerity, the unostentatious and deep-seated piety which breathes through all his correspondence, are in themselves a sufficient refutation of the hypothesis, invented by bigots to cover uncharitableness, that such opinions as his must arise from moral defects. And his statue will do as good service as the brazen image that was set upon a pole before the Israelites, if those who have been bitten by the fiery serpents of sectarian hatred, which still haunt this wilderness of a world, are made whole by looking upon the image of a heretic, who was yet a saint.

Though Priestley did not believe in the natural immortality of man, he held with an almost naïve realism, that man would be raised from the dead by a direct exertion of the power of God, and thenceforward be immortal. And it may be as well for those who may be shocked by this doctrine to know that views, substantially identical with Priestley's, have been advocated, since his time, by two prelates of the Anglican Church: by Dr. Whately, Archbishop of Dublin, in his well-known "Essays;"[1] and by Dr. Courtenay, Bishop of Kingston in Jamaica, the first edition of whose remarkable book "On the Future States," dedicated to Archbishop Whately, was published in 1843 and the second in 1857. According to Bishop Courtenay,

> "The death of the body will cause a cessation of all the activity of the mind by way of natural consequence; to continue for ever UNLESS the Creator should interfere."

And again:—

> "The natural end of human existence is the 'first death,' the dreamless slumber of the grave, wherein man lies spell-bound, soul and body, under the dominion of sin and death—that whatever modes of conscious existence, whatever future states of 'life' or of 'torment' beyond Hades are reserved for

---

[1] First Series. "On Some of the Peculiarities of the Christian Religion." Essay I. Revelation of a Future State.

man, are results of our blessed Lord's victory over sin and death; that the resurrection of the dead must be preliminary to their entrance into either of the future states, and that the nature and even existence of these states and even the mere fact that there is a futurity of consciousness, can be known *only* through God's revelation of Himself in the Person and the Gospel of His Son."—P. 389.

And now hear Priestley:—

"Man, according to this system (of materialism), is no more than we now see of him. His being commences at the time of his conception, or perhaps at an earlier period. The corporeal and mental faculties, in being in the same substance, grow, ripen, and decay together; and whenever the system is dissolved it continues in a state of dissolution till it shall please that Almighty Being who called it into existence to restore it to life again."—"Matter and Spirit," p. 49.

And again:—

"The doctrine of the Scripture is, that God made man of the dust of the ground, and by simply animating this organised matter, made man that living percipient and intelligent being that he is. According to Revelation, *death* is a state of rest and insensibility, and our only though sure hope of a future life is founded on the doctrine of the resurrection of the whole man at some distant period; this assurance being sufficiently confirmed to us both by the evident tokens of a Divine commission attending the persons who delivered the doctrine, and especially by the actual resurrection of Jesus Christ, which is more authentically attested than any other fact in history."—*Ibid.*, p. 247.

We all know that "a saint in crape is twice a saint in lawn;" but it is not yet admitted that the views which are consistent with such saintliness in lawn, become diabolical when held by a mere dissenter.[1]

[1] Not only is Priestley at one with Bishop Courtenay in this

I am not here either to defend or to attack Priestley's philosophical views, and I cannot say that I am personally disposed to attach much value to episcopal authority in philosophical questions; but it seems right to call attention to the fact, that those of Priestley's opinions which have brought most odium upon him, have been openly promulgated, without challenge, by persons occupying the highest positions in the State Church.

I must confess that what interests me most about Priestley's materialism, is the evidence that he saw dimly the seed of destruction which such materialism carries within its own bosom. In the course of his reading for his "History of Discoveries relating to Vision, Light, and Colours," he had come upon the speculations of Boscovich and Michell, and had been led to admit the sufficiently obvious truth that our knowledge of matter is a knowledge of its properties; and that of its substance—if it have a substance—we know nothing. And this led to the further admission that, so far as we can know, there may be no difference between the substance of matter and the substance of spirit ("Disquisitions," p. 16). A step

matter, but with Hartley and Bonnet, both of them stout champions of Christianity. Moreover, Archbishop Whately's essay is little better than an expansion of the first paragraph of Hume's famous essay on the Immortality of the Soul:—"By the mere light of reason it seems difficult to prove the immortality of the soul; the arguments for it are commonly derived either from metaphysical topics, or moral, or physical. But it is in reality the Gospel, and the Gospel alone, that has brought *life and immortality to light*." It is impossible to imagine that a man of Whately's tastes and acquirements had not read Hume or Hartley, though he refers to neither.

farther would have shown Priestley that his materialism was, essentially, very little different from the Idealism of his contemporary, the Bishop of Cloyne.

As Priestley's philosophy is mainly a clear statement of the views of the deeper thinkers of his day, so are his political conceptions based upon those of Locke. Locke's aphorism that "the end of government is the good of mankind," is thus expanded by Priestley:—

"It must necessarily be understood, therefore, whether it be expressed or not, that all people live in society for their mutual advantage; so that the good and happiness of the members, that is, of the majority of the members, of any state, is the great standard by which everything relating to that state must finally be determined."[1]

The little sentence here interpolated, "that is, of the majority of the members of any state," appears to be that passage which suggested to Bentham, according to his own acknowledgment, the famous "greatest happiness" formula, which by substituting "happiness" for "good," has converted a noble into an ignoble principle. But I do not call to mind that there is any utterance in Locke quite so outspoken as the following passage in the "Essay on the First Principles of Government." After laying down as "a fundamental maxim in all governments," the proposition that "kings, senators, and nobles" are "the servants of the public," Priestley goes on to say:—

[1] "Essay on the First Principles of Government." Second edition, 1771, p. 13.

"But in the largest states, if the abuses of the government should at any time be great and manifest; if the servants of the people, forgetting their masters and their masters' interest, should pursue a separate one of their own; if, instead of considering that they are made for the people, they should consider the people as made for them; if the oppressions and violation of right should be great, flagrant, and universally resented; if the tyrannical governors should have no friends but a few sycophants, who had long preyed upon the vitals of their fellow-citizens, and who might be expected to desert a government whenever their interests should be detached from it; if, in consequence of these circumstances, it should become manifest that the risk which would be run in attempting a revolution would be trifling, and the evils which might be apprehended from it were far less than those which were actually suffered and which were daily increasing; in the name of God, I ask, what principles are those which ought to restrain an injured and insulted people from asserting their natural rights, and from changing or even punishing their governors—that is, their servants—who had abused their trust, or from altering the whole form of their government, if it appeared to be of a structure so liable to abuse?"

As a Dissenter, subject to the operation of the Corporation and Test Acts, and as a Unitarian, excluded from the benefit of the Toleration Act, it is not surprising to find that Priestley had very definite opinions about Ecclesiastical Establishments; the only wonder is that these opinions were so moderate as the following passages show them to have been:—

"Ecclesiastical authority may have been necessary in the infant state of society, and, for the same reason, it may perhaps continue to be, in some degree, necessary as long as society is imperfect; and therefore may not be entirely abolished till civil governments have arrived at a much greater degree of perfection. If, therefore, I were asked whether I should approve of the immediate dissolution of all the ecclesiastical establishments in

Europe, I should answer, No. . . . Let experiment be first made of *alterations*, or, which is the same thing, of *better establishments* than the present. Let them be reformed in many essential articles, and then not thrown aside entirely till it be found by experience that no good can be made of them."

Priestley goes on to suggest four such reforms of a capital nature :—

"1. Let the Articles of Faith to be subscribed by candidates for the ministry be greatly reduced. In the formulary of the Church of England, might not thirty-eight out of the thirty-nine be very well spared? It is a reproach to any Christian establishment if every man cannot claim the benefit of it who can say that he believes in the religion of Jesus Christ as it is set forth in the New Testament. You say the terms are so general that even Deists would quibble and insinuate themselves. I answer that all the articles which are subscribed at present, by no means exclude Deists who will prevaricate; and upon this scheme you would at least exclude fewer honest men."[1]

The second reform suggested is the equalisation, in proportion to work done, of the stipends of the clergy; the third, the exclusion of the bishops from Parliament; and the fourth, complete toleration, so that every man may enjoy the rights of a citizen, and be qualified to serve his country, whether he belong to the Established Church or not.

Opinions such as those I have quoted, respecting the duties and the responsibilities of governors, are the commonplaces of modern Liberalism; and Priestley's views on Ecclesiastical Establishments would, I fear, meet with but a cool reception, as altogether too conservative, from a large proportion of the lineal

[1] "Utility of Establishments," in "Essay on First Principles of Government," p. 198, 1771.

descendants of the people who taught their children to cry "Damn Priestley;" and, with that love for the practical application of science which is the source of the greatness of Birmingham, tried to set fire to the doctor's house with sparks from his own electrical machine; thereby giving the man they called an incendiary and raiser of sedition against Church and King, an appropriately experimental illustration of the nature of arson and riot.

If I have succeeded in putting before you the main features of Priestley's work, its value will become apparent, when we compare the condition of the English nation, as he knew it, with its present state.

The fact that France has been for eighty-five years trying, without much success, to right herself after the great storm of the Revolution, is not unfrequently cited among us, as an indication of some inherent incapacity for self-government among the French people. I think, however, that Englishmen who argue thus, forget that, from the meeting of the Long Parliament in 1640, to the last Stuart rebellion in 1745, is a hundred and five years, and that, in the middle of the last century, we had but just safely freed ourselves from our Bourbons and all that they represented. The corruption of our state was as bad as that of the Second Empire. Bribery was the instrument of government, and peculation its reward. Four-fifths of the seats in the House of Commons were more or less openly dealt with as property. A minister had to consider the state of the vote market, and the sovereign secured a sufficiency of "king's friends" by payments allotted with retail, rather than royal, sagacity.

Barefaced and brutal immorality and intemperance pervaded the land, from the highest to the lowest classes of society. The Established Church was torpid, so far as it was not a scandal; but those who dissented from it came within the meshes of the Act of Uniformity, the Test Act, and the Corporation Act. By law, such a man as Priestley, being a Unitarian, could neither teach nor preach, and was liable to ruinous fines and long imprisonment.[1] In those days, the guns that were pointed by the Church against the Dissenters were shotted. The law was a cesspool of iniquity and cruelty. Adam Smith was a new prophet whom few regarded, and commerce was hampered by idiotic impediments, and ruined by still more absurd help, on the part of government.

Birmingham, though already the centre of a considerable industry, was a mere village as compared with its present extent. People who travelled went about armed, by reason of the abundance of highwaymen and the paucity and inefficiency of the police. Stage coaches had not reached Birmingham, and it took three days to get to London. Even canals were a recent and much opposed invention.

Newton had laid the foundation of a mechanical conception of the physical universe: Hartley, putting a

[1] In 1732 Doddridge was cited for teaching without the Bishop's leave, at Northampton.

modern face upon ancient materialism, had extended that mechanical conception to psychology; Linnæus and Haller were beginning to introduce method and order into the chaotic accumulation of biological facts. But those parts of physical science which deal with heat, electricity, and magnetism, and above all, chemistry, in the modern sense, can hardly be said to have had an existence. No one knew that two of the old elemental bodies, air and water, are compounds, and that a third, fire, is not a substance but a motion. The great industries that have grown out of the applications of modern scientific discoveries had no existence, and the man who should have foretold their coming into being in the days of his son, would have been regarded as a mad enthusiast.

In common with many other excellent persons, Priestley believed that man is capable of reaching, and will eventually attain, perfection. If the temperature of space presented no obstacle, I should be glad to entertain the same idea; but judging from the past progress of our species, I am afraid that the globe will have cooled down so far, before the advent of this natural millennium, that we shall be, at best, perfected Esquimaux. For all practical purposes, however, it is enough that man may visibly improve his condition in the course of a century or so. And, if the picture of the state of things in Priestley's time, which I have just drawn, have any pretence to accuracy, I think it must be admitted that there has been a considerable change for the better.

I need not advert to the well-worn topic of material

advancement, in a place in which the very stones testify to that progress—in the town of Watt and of Boulton. I will only remark, in passing, that material advancement has its share in moral and intellectual progress. Becky Sharp's acute remark that it is not difficult to be virtuous on ten thousand a year, has its application to nations; and it is futile to expect a hungry and squalid population to be anything but violent and gross. But as regards other than material welfare, although perfection is not yet in sight—even from the mast-head—it is surely true that things are much better than they were.

Take the upper and middle classes as a whole, and it may be said that open immorality and gross intemperance have vanished. Four and six bottle men are as extinct as the dodo. Women of good repute do not gamble, and talk modelled upon Dean Swift's "Art of Polite Conversation" would be tolerated in no decent kitchen.

Members of the legislature are not to be bought; and constituents are awakening to the fact that votes must not be sold—even for such trifles as rabbits and tea and cake. Political power has passed into the hands of the masses of the people. Those whom Priestley calls their servants have recognised their position, and have requested the master to be so good as to go to school and fit himself for the administration of his property. No civil disability attaches to any one on theological grounds, and the highest offices of the state are open to Papist, Jew, or Secularist.[1]

[1] The recent proceedings of the House of Commons throw a doubt,

Whatever men's opinions as to the policy of Establishment, no one can hesitate to admit that the clergy of the Church are men of pure life and conversation, zealous in the discharge of their duties; and, at present, apparently, more bent on prosecuting one another than on meddling with Dissenters. Theology itself has broadened so much, that Anglican divines put forward doctrines more liberal than those of Priestley; and, in our state-supported churches, one listener may hear a sermon to which Bossuet might have given his approbation, while another may hear a discourse in which Socrates would find nothing new.

But great as these changes may be, they sink into insignificance beside the progress of physical science, whether we consider the improvement of methods of investigation, or the increase in bulk of solid knowledge. Consider that the labours of Laplace, of Young, of Davy, and of Faraday; of Cuvier, of Lamarck, and of Robert Brown; of Von Baer, and of Schwann; of Smith and of Hutton, have all been carried on since Priestley discovered oxygen; and consider that they are now things of the past, concealed by the industry of those who have built upon them, as the first founders of a coral reef are hidden beneath the life's work of their successors; consider that the methods of physical science are slowly spreading into all investigations, and that proofs as valid as those required by her canons of investigation, are being demanded of all doctrines which ask for

which it is to be hoped may speedily be removed, on the accuracy of this statement. (September 1881.)

men's assent; and you will have a faint image of the astounding difference in this respect between the nineteenth century and the eighteenth.

If we ask what is the deeper meaning of all these vast changes, I think there can be but one reply. They mean that reason has asserted and exercised her primacy over all provinces of human activity: that ecclesiastical authority has been relegated to its proper place; that the good of the governed has been finally recognised as the end of government, and the complete responsibility of governors to the people as its means; and that the dependence of natural phenomena in general, on the laws of action of what we call matter has become an axiom.

But it was to bring these things about, and to enforce the recognition of these truths, that Joseph Priestley laboured. If the nineteenth century is other and better than the eighteenth, it is, in great measure, to him and to such men as he, that we owe the change. If the twentieth century is to be better than the nineteenth, it will be because there are among us men who walk in Priestley's footsteps.

Such men are not those whom their own generation delights to honour; such men, in fact, rarely trouble themselves about honour, but ask, in another spirit than Falstaff's, "What is honour? Who hath it? He that died o' Wednesday." But whether Priestley's lot be theirs, and a future generation, in justice and in gratitude, set up their statues; or whether their names and fame are blotted out from remembrance, their work will live as long as time endures. To all eternity, the sum of truth and right will have been increased by their means; to all eternity, falsehood and injustice will be the weaker because they have lived.

VI.

ON THE METHOD OF ZADIG:

RETROSPECTIVE PROPHECY AS A FUNCTION OF SCIENCE.

"Une marque plus sûre que toutes celles de Zadig"—CUVIER.[1]

IT is a usual and a commendable practice to preface the discussion of the views of a philosophic thinker by some account of the man and of the circumstances which shaped his life and coloured his way of looking at things; but, though Zadig is cited in one of the most important chapters of Cuvier's greatest work, little is known about him, and that little might perhaps be better authenticated than it is.

It is said that he lived at Babylon in the time of King Moabdar; but the name of Moabdar does not appear in the list of Babylonian sovereigns brought to light by the patience and the industry of the decipherers of cuneiform inscriptions in these later years; nor indeed am I aware that there is any other authority for his existence than that of the biographer of Zadig, one Arouet de Voltaire, among whose more

[1] "Discours sur les révolutions de la surface du globe," *Recherches sur les ossemens fossiles*, Ed. iv. t. i. p. 185.

conspicuous merits strict historical accuracy is perhaps hardly to be reckoned.

Happily Zadig is in the position of a great many other philosophers. What he was like when he was in the flesh, indeed whether he existed at all, are matters of no great consequence. What we care about in a light is that it shows the way, not whether it is lamp or candle, tallow or wax. Our only real interest in Zadig lies in the conceptions of which he is the putative father; and his biographer has stated these with so much clearness and vivacious illustration, that we need hardly feel a pang, even if critical research should prove King Moabdar and all the rest of the story to be unhistorical, and reduce Zadig himself to the shadowy condition of a solar myth.

Voltaire tells us that, disenchanted with life by sundry domestic misadventures, Zadig withdrew from the turmoil of Babylon to a secluded retreat on the banks of the Euphrates, where he beguiled his solitude by the study of nature. The manifold wonders of the world of life had a particular attraction for the lonely student; incessant and patient observation of the plants and animals about him sharpened his naturally good powers of observation and of reasoning; until, at length, he acquired a sagacity which enabled him to perceive endless minute differences among objects which, to the untutored eye, appeared absolutely alike.

It might have been expected that this enlargement of the powers of the mind and of its store of natural knowledge could tend to nothing but the increase of a man's own welfare and the good of his fellow-men.

But Zadig was fated to experience the vanity of such expectations.

One day, walking near a little wood, he saw, hastening that way, one of the Queen's chief eunuchs, followed by a troop of officials, who appeared to be in the greatest anxiety, running hither and thither like men distraught, in search of some lost treasure.

"Young man," cried the eunuch, "have you seen the Queen's dog?" Zadig answered modestly, "A bitch, I think, not a dog." "Quite right," replied the eunuch; and Zadig continued, "A very small spaniel who has lately had puppies; she limps with the left foreleg, and has very long ears." "Ah! you have seen her then," said the breathless eunuch. "No," answered Zadig, "I have not seen her; and I really was not aware that the Queen possessed a spaniel."

By an odd coincidence, at the very same time, the handsomest horse in the King's stables broke away from his groom in the Babylonian plains. The grand huntsman and all his staff were seeking the horse with as much anxiety as the eunuch and his people the spaniel; and the grand huntsman asked Zadig if he had not seen the King's horse go that way.

"A first-rate galloper, small-hoofed, five feet high; tail three feet and a half long; cheek pieces of the bit of twenty-three carat gold; shoes silver?" said Zadig.

"Which way did he go? Where is he?" cried the grand huntsman.

"I have not seen anything of the horse, and I never heard of him before," replied Zadig.

The grand huntsman and the chief eunuch made sure that Zadig had stolen both the King's horse and the Queen's spaniel, so they haled him before the High Court of Desterham, which at once condemned him to the knout, and transportation for life to Siberia. But the sentence was hardly pronounced when the lost horse and spaniel were found. So the judges were under the painful necessity of reconsidering their decision: but they fined Zadig four hundred ounces of gold for saying he had seen that which he had not seen.

The first thing was to pay the fine; afterwards Zadig was

permitted to open his defence to the court, which he did in the following terms:

"Stars of justice, abysses of knowledge, mirrors of truth, whose gravity is as that of lead, whose inflexibility is as that of iron, who rival the diamond in clearness, and possess no little affinity with gold; since I am permitted to address your august assembly, I swear by Ormuzd that I have never seen the respectable lady dog of the queen, nor beheld the sacrosanct horse of the King of Kings.

"This is what happened. I was taking a walk towards the little wood near which I subsequently had the honour to meet the venerable chief eunuch and the most illustrious grand huntsman. I noticed the track of an animal in the sand, and it was easy to see that it was that of a small dog. Long faint streaks upon the little elevations of sand between the footmarks convinced me that it was a she dog with pendent dugs, showing that she must have had puppies not many days since. Other scrapings of the sand, which always lay close to the marks of the forepaws, indicated that she had very long ears; and, as the imprint of one foot was always fainter than those of the other three, I judged that the lady dog of our august Queen was, if I may venture to say so, a little lame.

"With respect to the horse of the King of Kings, permit me to observe that, wandering through the paths which traverse the wood, I noticed the marks of horse-shoes. They were all equidistant. 'Ah!' said I, 'this is a famous galloper.' In a narrow alley, only seven feet wide, the dust upon the trunks of the trees was a little disturbed at three feet and a half from the middle of the path. 'This horse,' said I to myself, 'had a tail three feet and a half long, and, lashing it from one side to the other, he has swept away the dust.' Branches of the trees met overhead at the height of five feet, and under them I saw newly fallen leaves; so I knew that the horse had brushed some of the branches, and was therefore five feet high. As to his bit, it must have been made of twenty-three carat gold, for he had rubbed it against a stone, which turned out to be a touchstone, with the properties of which I am familiar by experiment. Lastly, by the marks which his shoes left upon pebbles of another kind, I was led to think that his shoes were of fine silver."

All the judges admired Zadig's profound and subtle discernment; and the fame of it reached even the King and the Queen. From the ante-rooms to the presence-chamber, Zadig's name was in everybody's mouth; and, although many of the magi were of opinion that he ought to be burnt as a sorcerer, the King commanded that the four hundred ounces of gold which he had been fined should be restored to him. So the officers of the court went in state with the four hundred ounces; only they retained three hundred and ninety-eight for legal expenses, and their servants expected fees.

Those who are interested in learning more of the fateful history of Zadig must turn to the original; we are dealing with him only as a philosopher, and this brief excerpt suffices for the exemplification of the nature of his conclusions and of the method by which he arrived at them.

These conclusions may be said to be of the nature of retrospective prophecies; though it is perhaps a little hazardous to employ phraseology which perilously suggests a contradiction in terms—the word "prophecy" being so constantly in ordinary use restricted to "foretelling." Strictly, however, the term prophecy as much applies to outspeaking as to foretelling; and, even in the restricted sense of "divination," it is obvious that the essence of the prophetic operation does not lie in its backward or forward relation to the course of time, but in the fact that it is the apprehension of that which lies out of the sphere of immediate knowledge; the seeing of that which to the natural sense of the seer is invisible.

The foreteller asserts that, at some future time, a properly situated observer will witness certain events;

the clairvoyant declares that, at this present time, certain things are to be witnessed a thousand miles away; the retrospective prophet (would that there were such a word as "backteller!") affirms that so many hours or years ago, such and such things were to be seen. In all these cases, it is only the relation to time which alters—the process of divination beyond the limits of possible direct knowledge remains the same.

No doubt it was their instinctive recognition of the analogy between Zadig's results and those obtained by authorised inspiration which inspired the Babylonian magi with the desire to burn the philosopher. Zadig admitted that he had never either seen or heard of the horse of the king or of the spaniel of the queen; and yet he ventured to assert in the most positive manner that animals answering to their description did actually exist, and ran about the plains of Babylon. If his method was good for the divination of the course of events ten hours old, why should it not be good for those of ten years or ten centuries past; nay, might it not extend to ten thousand years and justify the impious in meddling with the traditions of Oannes and the fish, and all the sacred foundations of Babylonian cosmogony?

But this was not the worst. There was another consideration which obviously dictated to the more thoughtful of the magi the propriety of burning Zadig out of hand. His defence was worse than his offence. It showed that his mode of divination was fraught with danger to magianism in general. Swollen with the pride of human reason, he had ignored the established canons of magian lore; and, trusting to what after all was mere carnal common sense, he professed to lead men to a deeper insight into nature than magian wisdom, with all its lofty antagonism to everything common, had ever reached. What, in fact, lay at the foundation of all Zadig's arguments but the coarse commonplace assumption, upon which every act of our daily lives is based, that we may conclude from an effect to the pre-existence of a cause competent to produce that effect?

The tracks were exactly like those which dogs and horses leave; therefore they were the effects of such animals as causes. The marks at the sides of the fore prints of the dog track were exactly such as would be produced by long trailing ears; therefore the dog's long ears were the causes of these marks—and so on. Nothing can be more hopelessly vulgar, more unlike the majestic development of a system of grandly unintelligible conclusions from sublimely inconceivable premisses, such as delights the magian heart. In fact, Zadig's method was nothing but the method of all mankind. Retrospective prophecies, far more astonishing for their minute accuracy than those of Zadig, are familiar to those who have watched the daily life of nomadic people.

From freshly broken twigs, crushed leaves, disturbed pebbles, and imprints hardly discernible by the untrained eye, such graduates in the University of Nature will divine, not only the fact that a party has passed that way, but its strength, its composition, the course it took, and the number of hours or days which have elapsed since it passed. But they are able to do this because, like Zadig, they perceive endless minute differences where untrained eyes discern nothing; and because the unconscious logic of common sense compels them to account for these effects by the causes which they know to be competent to produce them.

And such mere methodised savagery was to discover the hidden things of nature better than *à priori* deductions from the nature of Ormuzd—perhaps to give a history of the past, in which Oannes would be altogether ignored! Decidedly it were better to burn this man at once.

If instinct, or an unwonted use of reason, led Moabdar's magi to this conclusion two or three thousand years ago, all that can be said is that subsequent history has fully justified them. For the rigorous application of Zadig's logic to the results of accurate and long-continued observation has founded all those sciences which have been termed historical or palætiological, because they are retrospectively prophetic and strive towards the reconstruction in human imagination of events which have vanished and ceased to be.

History, in the ordinary acceptation of the word, is based upon the interpretation of documentary evidence; and documents would have no evidential value unless historians were justified in their assumption that they have come into existence by the operation of causes similar to those of which documents are, in our present experience, the effects. If a written history can be produced otherwise than by human agency, or if the man who wrote a given document was actuated by other than ordinary human motives, such documents are of no more evidential value than so many arabesques.

Archæology, which takes up the thread of history beyond the point at which documentary evidence fails us, could have no existence, except for our well-grounded confidence that monuments and works of art or artifice, have never been produced by causes different in kind from those to which they now owe their origin. And geology, which traces back the course of history beyond the limits of archæology, could tell us nothing except for the assumption that, millions of years ago, water, heat, gravitation, friction, animal and vegetable life, caused effects of the same kind as they do now. Nay, even physical astronomy, in so far as it takes us back to the uttermost point of time which palætiological science can reach, is founded upon the same assumption. If the law of gravitation ever failed to be true, even to the smallest extent, for that period, the calculations of the astronomer have no application.

The power of prediction, of prospective prophecy, is that which is commonly regarded as the great prerogative of physical science. And truly it is a wonderful fact that one can go into a shop and buy for small price a book, the "Nautical Almanac," which will foretell the exact position to be occupied by one of Jupiter's moons six months hence; nay more, that, if it were worth while, the Astronomer Royal could furnish us with as infallible a prediction applicable to 1980 or 2980.

But astronomy is not less remarkable for its power of retrospective prophecy.

Thales, oldest of Greek philosophers, the dates of whose birth and death are uncertain, but who flourished about 600 B.C., is said to have foretold an eclipse of the sun which took place in his time during a battle between the Medes and the Lydians. Sir George Airy has written a very learned and interesting memoir[1] in which he proves that such an eclipse was visible in Lydia on the afternoon of the 28th of May in the year 585 B.C.

No one doubts that, on the day and at the hour mentioned by the Astronomer Royal, the people of Asia Minor saw the face of the sun totally obscured. But, though we implicitly believe this retrospective prophecy, it is incapable of verification. In the total absence of historical records, it is impossible even to conceive any means of ascertaining directly whether the eclipse of Thales happened or not. All that can be said is, that the prospective prophecies of the astronomer are always verified; and that, inasmuch as his retrospective prophecies are the result of following backwards, the very same method as that which invariably leads to verified results, when it is worked forwards, there is as much reason for placing full confidence in the one as in the other. Retrospective prophecy is therefore a legitimate function of astronomical science; and if it is legitimate for one science it is legitimate for all; the fundamental

[1] "On the Eclipses of Agathocles, Thales, and Xerxes," *Philosophical Transactions*, vol. cxliii.

axiom on which it rests, the constancy of the order of nature, being the common foundation of all scientific thought. Indeed, if there can be grades in legitimacy, certain branches of science have the advantage over astronomy, in so far as their retrospective prophecies are not only susceptible of verification, but are sometimes strikingly verified.

Such a science exists in that application of the principles of biology to the interpretation of the animal and vegetable remains imbedded in the rocks which compose the surface of the globe, which is called Palæontology.

At no very distant time, the question whether these so-called "fossils" were really the remains of animals and plants was hotly disputed. Very learned persons maintained that they were nothing of the kind, but a sort of concretion, or crystallisation, which had taken place within the stone in which they are found; and which simulated the forms of animal and vegetable life, just as frost on a window-pane imitates vegetation. At the present day, it would probably be impossible to find any sane advocate of this opinion; and the fact is rather surprising, that among the people from whom the circle-squarers, perpetual-motioners, flat-earth men and the like, are recruited, to say nothing of table-turners and spirit-rappers, somebody has not perceived the easy avenue to nonsensical notoriety open to any one who will take up the good old doctrine, that fossils are all *lusus naturæ*.

The position would be impregnable, inasmuch as it is quite impossible to prove the contrary. If a man choose to maintain that a fossil oyster shell, in spite of its correspondence, down to every minutest particular, with that of an oyster fresh taken out of the sea, was never tenanted by a living oyster, but is a mineral concretion, there is no demonstrating his error. All that can be done is to show him that, by a parity of reasoning, he is bound to admit that a heap of oyster shells outside a fishmonger's door may also be "sports of nature," and that a mutton bone in a dust-bin may have had the like origin. And when you cannot prove that people are wrong, but only that they are absurd, the best course is to let them alone.

The whole fabric of palæontology, in fact, falls to the ground unless we admit the validity of Zadig's great principle, that like effects imply like causes; and that the process of reasoning from a shell, or a tooth, or a bone, to the nature of the animal to which it belonged, rests absolutely on the assumption that the likeness of this shell, or tooth, or bone, to that of some animal with which we are already acquainted, is such that we are justified in inferring a corresponding degree of likeness in the rest of the two organisms. It is on this very simple principle, and not upon imaginary laws of physiological correlation, about which, in most cases, we know nothing whatever, that the so-called restorations of the palæontologist are based.

Abundant illustrations of this truth will occur to every one who is familiar with palæontology; none is more suitable than the case of the so-called *Belemnites*. In the early days of the study of fossils, this name was given to certain elongated stony bodies, ending at one extremity in a conical point, and truncated at the other, which were commonly reputed to be thunderbolts, and as such to have descended from the sky. They are common enough in some parts of England; and, in the condition in which they are ordinarily found, it might be difficult to give satisfactory reasons for denying them to be merely mineral bodies.

They appear, in fact, to consist of nothing but concentric layers of carbonate of lime, disposed in subcrystalline fibres, or prisms, perpendicular to the layers. Among a great number of specimens of these Belemnites, however, it was soon observed that some showed a conical cavity at the blunt end; and, in still better preserved specimens, this cavity appeared to be divided into chambers by delicate saucer-shaped partitions, situated at regular intervals one above the other. Now there is no mineral body which presents any structure comparable to this, and the conclusion suggested itself that the Belemnites must be the effects of causes other than those which are at work in inorganic nature. On close examination, the saucer-shaped partitions were proved to be all perforated at one point, and the perforations being situated exactly in the same line, the chambers were seen to be traversed by a canal, or *siphuncle*, which thus connected the smallest or apical chamber with the largest. There is nothing like this in the vegetable world; but an exactly corresponding structure is met with in the shells of two kinds of existing animals, the pearly *Nautilus* and the *Spirula*, and only in them. These animals belong to the same division—the *Cephalopoda* —as the cuttle-fish, the squid, and the octopus. But they are the only existing members of the group which possess chambered, siphunculated shells; and it is utterly impossible to trace any physiological connection between the very peculiar structural characters of a cephalopod and the presence of a chambered shell. In fact, the squid has, instead of any such shell, a horny "pen," the cuttle-fish has the so-called "cuttle-bone," and the octopus has no shell, or, at most, a mere rudiment of one.

Nevertheless, seeing that there is nothing in nature at all like the chambered shell of the Belemnite, except the shells of the *Nautilus* and of the *Spirula*, it was legitimate to prophesy that the animal from which the fossil proceeded must have belonged to the group of the *Cephalopoda*. *Nautilus* and *Spirula* are both very rare animals, but the progress of investigation brought to light the singular fact, that, though each has the characteristic cephalopodous organisation, it is very different from the other. The shell of *Nautilus* is external, that of *Spirula* internal; *Nautilus* has four gills, *Spirula* two; *Nautilus* has multitudinous tentacles, *Spirula* has only ten arms beset with horny rimmed suckers; *Spirula*, like the squids and cuttle-fishes, which it closely resembles, has a bag of ink which it squirts out to cover its retreat when alarmed; *Nautilus* has none.

No amount of physiological reasoning could enable any one to say whether the animal which fabricated the Belemnite was more like *Nautilus*, or more like

Spirula. But the accidental discovery of Belemnites in due connection with black elongated masses which were certainly fossilised ink-bags, inasmuch as the ink could be ground up and used for painting as well as if it were recent sepia, settled the question; and it became perfectly safe to prophesy that the creature which fabricated the Belemnite was a two-gilled cephalopod with suckers on its arms, and with all the other essential features of our living squids, cuttle-fishes, and *Spirulæ.* The palæontologist was, by this time, able to speak as confidently about the animal of the Belemnite, as Zadig was respecting the queen's spaniel. He could give a very fair description of its external appearance, and even enter pretty fully into the details of its internal organisation, and yet could declare that neither he, nor any one else, had ever seen one. And as the queen's spaniel was found, so happily has the animal of the Belemnite; a few exceptionally preserved specimens having been discovered, which completely verify the retrospective prophecy of those who interpreted the facts of the case by due application of the method of Zadig.

These Belemnites flourished in prodigious abundance in the seas of the mesozoic or secondary age of the world's geological history; but no trace of them has been found in any of the tertiary deposits, and they appear to have died out towards the close of the mesozoic epoch. The method of Zadig, therefore, applies in full force to the events of a period which is immeasurably remote, which long preceded the origin of the most conspicuous mountain masses of the present world, and the deposition, at the bottom of the ocean, of the rocks which form the greater part of the soil of our present continents. The Euphrates itself, at the mouth of which Oannes landed, is a thing of yesterday compared with a Belemnite; and even the liberal chronology of Magian cosmogony fixes the beginning of the world only at a time when other applications of Zadig's method afford convincing evidence that, could we have been there to see, things would have looked very much as they do now. Truly the magi were wise in their generation; they foresaw rightly that this pestilent application of the principles of common sense, inaugurated by Zadig, would be their ruin.

But it may be said that the method of Zadig, which is simple reasoning from analogy, does not account for the most striking feats of modern palæontology—the reconstruction of entire animals from a tooth or perhaps a fragment of a bone; and it may be justly urged that Cuvier, the great master of this kind of investigation, gave a very different account of the process which yielded such remarkable results.

Cuvier is not the first man of ability who has failed to make his own mental processes clear to himself, and he will not be the last. The matter can be easily tested. Search the eight volumes of the "Recherches sur les Ossemens fossiles" from cover to cover, and nothing but the application of the method of Zadig will be found in the arguments by which a fragment of a skeleton is made to reveal the characters of the animal to which it belonged.

There is one well-known case which may represent all. It is an excellent illustration of Cuvier's sagacity, and he evidently takes some pride in telling his story about it. A split slab of stone arrived from the quarries of Montmartre, the two halves of which contained the greater part of the skeleton of a small animal. On careful examinations of the characters of the teeth and of the lower jaw, which happened to be exposed, Cuvier assured himself that they presented such a very close resemblance to the corresponding parts in the living opossums that he at once assigned the fossil to that genus.

Now the opossums are unlike most mammals in that they possess two bones attached to the fore part of the pelvis, which are commonly called "marsupial bones." The name is a misnomer, originally conferred because it was thought that these bones have something to do with the support of the pouch, or marsupium, with which some, but not all, of the opossums are provided. As a matter of fact, they have nothing to do with the support of the pouch, and they exist as much in those opossums which have no pouches as in those which possess them. In truth, no one knows what the use of these bones may be, nor has any valid theory of their physiological import yet been suggested. And if we have no knowledge of the physiological importance of the bones themselves, it is obviously absurd to pretend that we are able to give physiological reasons why the presence of these bones is associated with certain peculiarities of the teeth and of the jaws. If any one knows why four molar teeth and an inflected angle of the jaw are very generally found along with marsupial bones, he has not yet communicated that knowledge to the world.

If, however, Zadig was right in concluding from the likeness of the hoof-prints which he observed to a horse's that the creature which made them had a tail like that of a horse, Cuvier, seeing that the teeth and jaw of his fossil were just like those of an opossum, had the same right to conclude that the pelvis would also be like an opossum's; and so strong was his conviction that this retrospective prophecy, about an animal which he had never seen before, and which had been dead and buried for millions of years, would be verified, that he went to work upon the slab which contained the pelvis in confident expectation of finding and laying bare the "marsupial bones," to the satisfaction of some persons whom he had invited to witness their disinterment. As he says:—"Cette opération se fit en présence de quelques personnes à qui j'en avais annoncé d'avance le résultat, dans l'intention de leur prouver par le fait la justice de nos théories zoologiques; puisque le vrai cachet d'une théorie est sans contredit la faculté qu'elle donne de prévoir les phénomènes."

In the "Ossemens fossiles" Cuvier leaves his paper just as it first appeared in the "Annales du Muséum," as "a curious monument of the force of zoological laws and of the use which may be made of them."

Zoological laws truly, but not physiological laws. If one sees a live dog's head, it is extremely probable that a dog's tail is not far off, though nobody can say why that sort of head and that sort of tail go together; what physiological connection there is between the two. So, in the case of the Montmartre fossil, Cuvier, finding a thorough opossum's head, concluded that the pelvis also would be like an opossum's. But, most assuredly, the most advanced physiologist of the present day could throw no light on the question why these are associated, nor could pretend to affirm that the existence of the one is necessarily connected with that of the other. In fact, had it so happened that the pelvis of the fossil had been originally exposed, while the head lay hidden, the presence of the "marsupial bones," however like they might have been to an opossum's, would by no means have warranted the prediction that the skull would turn out to be that of the opossum. It might just as well have been like that of some other Marsupial; or even like that of the totally different group of Monotremes, of which the only living representatives are the *Echidna* and the *Ornithorhynchus*.

For all practical purposes, however, the empirical laws of co-ordination of structures, which are embodied in the generalisations of morphology, may be confidently trusted, if employed with due caution, to lead to a just interpretation of fossil remains; or, in other words, we may look for the verification of the retrospective prophecies which are based upon them.

And if this be the case, the late advances which have been made in palæontological discovery open out a new field for such prophecies. For it has been ascertained with respect to many groups of animals, that, as we trace them back in time, their ancestors gradually cease to exhibit those special modifications which at present characterise the type, and more nearly embody the general plan of the group to which they belong.

Thus, in the well-known case of the horse, the toes which are suppressed in the living horse are found to be more and more complete in the older members of the group, until, at the bottom of the Tertiary series of America, we find an equine animal which has four toes in front and three behind. No remains of the horse tribe are at present known from any Mesozoic deposit. Yet who can doubt that, whenever a sufficiently extensive series of lacustrine and fluviatile beds of that age becomes known, the lineage which has been traced thus far will be continued by equine quadrupeds with an increasing number of digits, until the horse type merges in the five-toed form towards which these gradations point?

But the argument which holds good for the horse, holds good, not only for all mammals, but for the whole animal world. And as the study of the pedigrees, or lines of evolution, to which, at present, we have access, brings to light, as it assuredly will do, the laws of that process, we shall be able to reason from the facts with which the geological record furnishes us to those which have hitherto remained, and many of which, perhaps, may for ever remain, hidden. The same method of reasoning which enables us, when furnished with a fragment of an extinct animal, to prophesy the character which the whole organism exhibited, will, sooner or later, enable us, when we know a few of the later terms of a genealogical series, to predict the nature of the earlier terms.

In no very distant future, the method of Zadig, applied to a greater body of facts than the present generation is fortunate enough to handle, will enable the biologist to reconstruct the scheme of life from its beginning, and to speak as confidently of the character of long extinct living beings, no trace of which has been preserved, as Zadig did of the queen's spaniel and the king's horse. Let us hope that they may be better rewarded for their toil and their sagacity than was the Babylonian philosopher; for perhaps, by that time, the Magi also may be reckoned among the members of a forgotten Fauna, extinguished in the struggle for existence against their great rival, common sense.

VII.

ON THE BORDER TERRITORY BETWEEN THE ANIMAL AND THE VEGETABLE KINGDOMS.

In the whole history of science there is nothing more remarkable than the rapidity of the growth of biological knowledge within the last half-century, and the extent of the modification which has thereby been effected in some of the fundamental conceptions of the naturalist.

In the second edition of the "Règne Animal," published in 1828, Cuvier devotes a special section to the "Division of Organised Beings into Animals and Vegetables," in which the question is treated with that comprehensiveness of knowledge and clear critical judgment which characterise his writings, and justify us in regarding them as representative expressions of the most extensive, if not the profoundest, knowledge of his time. He tells us that living beings have been subdivided from the earliest times into *animated beings*, which possess sense and motion, and *inanimated beings*, which are devoid of these functions, and simply vegetate.

Although the roots of plants direct themselves towards moisture, and their leaves towards air and light,—although the parts of some plants exhibit oscillating movements without any perceptible cause, and the leaves of others retract when touched,—yet none of these movements justify the ascription to plants of perception or of will. From the mobility of animals, Cuvier, with his characteristic partiality for teleological reasoning, deduces the necessity of the existence in them of an alimentary cavity, or reservoir of food, whence their nutrition may be drawn by the vessels, which are a sort of internal roots; and, in the presence of this alimentary cavity, he naturally sees the primary and the most important distinction between animals and plants.

Following out his teleological argument, Cuvier remarks that the organisation of this cavity and its appurtenances must needs vary according to the nature of the aliment, and the operations which it has to undergo, before it can be converted into substances fitted for absorption; while the atmosphere and the earth supply plants with juices ready prepared, and which can be absorbed immediately. As the animal body required to be independent of heat and of the atmosphere, there were no means by which the motion of its fluids could be produced by internal causes. Hence arose the second great distinctive character of animals, or the circulatory system, which is less important than the digestive, since it was unnecessary, and therefore is absent, in the more simple animals.

Animals further needed muscles for locomotion and nerves for sensibility. Hence, says Cuvier, it was necessary that the chemical composition of the animal body should be more complicated than that of the plant; and it is so, inasmuch as an additional substance, nitrogen, enters into it as an essential element; while, in plants, nitrogen is only accidentally joined with the three other fundamental constituents of organic beings—carbon, hydrogen, and oxygen. Indeed, he afterwards affirms that nitrogen is peculiar to animals; and herein he places the third distinction between the animal and the plant. The soil and the atmosphere supply plants with water, composed of hydrogen and oxygen; air, consisting of nitrogen and oxygen; and carbonic acid, containing carbon and oxygen. They retain the hydrogen and the carbon, exhale the superfluous oxygen, and absorb little or no nitrogen. The essential character of vegetable life is the exhalation of oxygen, which is effected through the agency of light. Animals, on the contrary, derive their nourishment either directly or indirectly from plants. They get rid of the superfluous hydrogen and carbon, and accumulate nitrogen. The relations of plants and animals to the atmosphere are therefore inverse. The plant withdraws water and carbonic acid from the atmosphere, the animal contributes both to it. Respiration—that is, the absorption of oxygen and the exhalation of carbonic acid — is the specially animal function of animals, and constitutes their fourth distinctive character.

Thus wrote Cuvier in 1828. But, in the fourth and fifth decades of this century, the greatest and most rapid revolution which biological science has ever undergone was effected by the application of the modern microscope to the investigation of organic structure; by the introduction of exact and easily manageable methods of conducting the chemical analysis of organic compounds; and finally, by the employment of instruments of precision for the measurement of the physical forces which are at work in the living economy.

That the semi-fluid contents (which we now term protoplasm) of the cells of certain plants, such as the *Charæ*, are in constant and regular motion, was made out by Bonaventura Corti a century ago; but the fact, important as it was, fell into oblivion, and had to be rediscovered by Treviranus in 1807. Robert Brown noted the more complex motions of the protoplasm in the cells of *Tradescantia* in 1831; and now such movements of the living substance of plants are well known to be some of the most widely-prevalent phenomena of vegetable life.

Agardh, and other of the botanists of Cuvier's generation, who occupied themselves with the lower plants, had observed that, under particular circumstances, the contents of the cells of certain water-weeds were set free, and moved about with considerable velocity, and with all the appearances of spontaneity, as locomotive bodies, which, from their similarity to animals of simple organisation, were called "zoospores." Even as late as 1845, however, a botanist of Schleiden's eminence dealt very sceptically with these statements; and his scepticism was the more justified, since Ehrenberg, in his elaborate and comprehensive work on the *Infusoria*, had declared the greater number of what are now recognised as locomotive plants to be animals.

At the present day, innumerable plants and free plant cells are known to pass the whole or part of their lives in an actively locomotive condition, in no wise distinguishable from that of one of the simpler animals; and, while in this condition, their movements are, to all appearance, as spontaneous—as much the product of volition—as those of such animals.

Hence the teleological argument for Cuvier's first diagnostic character—the presence in animals of an alimentary cavity, or internal pocket, in which they can carry about their nutriment—has broken down, so far, at least, as his mode of stating it goes. And, with the advance of microscopic anatomy, the universality of the fact itself among animals has ceased to be predicable. Many animals of even complex structure, which live parasitically within others, are wholly devoid of an alimentary cavity. Their food is provided for them, not only ready cooked, but ready digested, and the alimentary canal, become superfluous, has disappeared. Again, the males of most Rotifers have no digestive apparatus; as a German naturalist has remarked, they devote themselves entirely to the "Minnedienst," and are to be reckoned among the few realisations of the Byronic ideal of a lover. Finally, amidst the lowest forms of animal life, the speck of gelatinous protoplasm, which constitutes the whole body, has no permanent digestive cavity or mouth, but takes in its food anywhere; and digests, so to speak, all over its body.

But although Cuvier's leading diagnosis of the animal from the plant will not stand a strict test, it remains one of the most constant of the distinctive characters of animals. And, if we substitute for the possession of an alimentary cavity, the power of taking solid nutriment into the body and there digesting it, the definition so changed will cover all animals, except certain parasites, and the few and exceptional cases of non-parasitic animals which do not feed at all. On the other hand, the definition thus amended will exclude all ordinary vegetable organisms.

Cuvier himself practically gives up his second distinctive mark when he admits that it is wanting in the simpler animals.

The third distinction is based on a completely erroneous conception of the chemical differences and resemblances between the constituents of animal and vegetable organisms, for which Cuvier is not responsible, as it was current among contemporary chemists. It is now established that nitrogen is as essential a constituent of vegetable as of animal living matter; and that the latter is, chemically speaking, just as complicated as the former. Starchy substances, cellulose and sugar, once supposed to be exclusively confined to plants, are now known to be regular and normal products of animals. Amylaceous and saccharine substances are largely manufactured, even by the highest animals; cellulose is widespread as a constituent of the skeletons of the lower animals; and it is probable that amyloid substances are universally present in the animal organism, though not in the precise form of starch.

Moreover, although it remains true that there is an inverse relation between the green plant in sunshine and the animal, in so far as, under these circumstances, the green plant decomposes carbonic acid and exhales oxygen, while the animal absorbs oxygen and exhales carbonic acid; yet, the exact researches of the modern chemical investigators of the physiological processes of plants have clearly demonstrated the fallacy of attempting to draw any general distinction between animals and vegetables on this ground. In fact, the difference vanishes with the sunshine, even in the case of the green plant; which, in the dark, absorbs oxygen and gives out carbonic acid like any animal.[1] On the other hand, those plants, such as the fungi, which contain no chlorophyll and are not green, are always, so far as respiration is concerned, in the exact position of animals. They absorb oxygen and give out carbonic acid.

Thus, by the progress of knowledge, Cuvier's fourth distinction between the animal and the plant has been as completely invalidated as the third and

[1] There is every reason to believe that living plants, like living animals, always respire, and, in respiring, absorb oxygen and give off carbonic acid; but, that in green plants exposed to daylight or to the electric light, the quantity of oxygen evolved in consequence of the decomposition of carbonic acid by a special apparatus which green plants possess exceeds that absorbed in the concurrent respiratory process.

second; and even the first can be retained only in a modified form and subject to exceptions.

But has the advance of biology simply tended to break down old distinctions, without establishing new ones?

With a qualification, to be considered presently, the answer to this question is undoubtedly in the affirmative. The famous researches of Schwann and Schleiden in 1837 and the following years, founded the modern science of histology, or that branch of anatomy which deals with the ultimate visible structure of organisms, as revealed by the microscope; and, from that day to this, the rapid improvement of methods of investigation, and the energy of a host of accurate observers, have given greater and greater breadth and firmness to Schwann's great generalisation, that a fundamental unity of structure obtains in animals and plants; and that, however diverse may be the fabrics, or *tissues*, of which their bodies are composed, all these varied structures result from the metamorphosis of morphological units (termed *cells*, in a more general sense than that in which the word "cells" was at first employed), which are not only similar in animals and in plants respectively, but present a close resemblance, when those of animals and those of plants are compared together.

The contractility which is the fundamental condition of locomotion, has not only been discovered to exist far more widely among plants than was formerly imagined; but, in plants, the act of contraction has been found to be accompanied, as Dr. Burdon Sanderson's interesting investigations have shown, by a disturbance of the electrical state of the contractile substance, comparable to that which was found by Du Bois Reymond to be a concomitant of the activity of ordinary muscle in animals.

Again, I know of no test by which the reaction of the leaves of the Sundew and of other plants to stimuli, so fully and carefully studied by Mr. Darwin, can be distinguished from those acts of contraction following upon stimuli, which are called "reflex" in animals.

On each lobe of the bilobed leaf of Venus's fly trap (*Dionæa muscipula*) are three delicate filaments which stand out at right angle from the surface of the leaf. Touch one of them with the end of a fine human hair and the lobes of the leaf instantly close together[1] in virtue of an act of contraction of part of their substance, just as the body of a snail contracts into its shell when one of its "horns" is irritated.

The reflex action of the snail is the result of the presence of a nervous system in the animal. A molecular change takes place in the nerve of the tentacle, is propagated to the muscles by which the body is retracted, and causing them to contract, the act of retraction is brought about. Of course the similarity of the acts does not necessarily involve the conclusion that the mechanism by which they are effected is the same; but it suggests a suspicion of their identity which needs careful testing.

The results of recent inquiries into the structure of

[1] Darwin. "Insectivorous Plants," p. 289.

the nervous system of animals converge towards the conclusion that the nerve fibres, which we have hitherto regarded as ultimate elements of nervous tissue, are not such, but are simply the visible aggregations of vastly more attenuated filaments, the diameter of which dwindles down to the limits of our present microscopic vision, greatly as these have been extended by modern improvements of the microscope; and that a nerve is, in its essence, nothing but a linear tract of specially modified protoplasm between two points of an organism—one of which is able to affect the other by means of the communication so established. Hence, it is conceivable that even the simplest living being may possess a nervous system. And the question whether plants are provided with a nervous system or not, thus acquires a new aspect, and presents the histologist and physiologist with a problem of extreme difficulty, which must be attacked from a new point of view and by the aid of methods which have yet to be invented.

Thus it must be admitted that plants may be contractile and locomotive; that, while locomotive, their movements may have as much appearance of spontaneity as those of the lowest animals; and that many exhibit actions, comparable to those which are brought about by the agency of a nervous system in animals. And it must be allowed to be possible that further research may reveal the existence of something comparable to a nervous system in plants. So that I know not where we can hope to find any absolute distinction between animals and plants, unless we return to their mode of nutrition, and inquire whether certain differences of a more occult character than those imagined to exist by Cuvier, and which certainly hold good for the vast majority of animals and plants, are of universal application.

A bean may be supplied with water in which salts of ammonia and certain other mineral salts are dissolved in due proportion; with atmospheric air containing its ordinary minute dose of carbonic acid; and with nothing else but sunlight and heat. Under these circumstances, unnatural as they are, with proper management, the bean will thrust forth its radicle and its plumule; the former will grow down into roots, the latter grow up into the stem and leaves of a vigorous bean plant; and this plant will, in due time, flower and produce its crop of beans, just as if it were grown in the garden or in the field.

The weight of the nitrogenous protein compounds, of the oily, starchy, saccharine and woody substances contained in the full-grown plant and its seeds, will be vastly greater than the weight of the same substances contained in the bean from which it sprang. But nothing has been supplied to the bean save water, carbonic acid, ammonia, potash, lime, iron, and the like, in combination with phosphoric, sulphuric, and other acids. Neither protein, nor fat, nor starch, nor sugar, nor any substance in the slightest degree resembling them, has formed part of the food of the bean. But the weights of the carbon, hydrogen, oxygen, nitrogen, phosphorus, sulphur, and other elementary bodies contained in the bean-plant, and in the seeds which it produces, are exactly equivalent to the weights of the same elements which have disappeared from the materials supplied to the bean during its growth. Whence it follows that the bean has taken in only the raw materials of its fabric, and has manufactured them into bean stuffs.

The bean has been able to perform this great chemical feat by the help of its green colouring matter, or chlorophyll; for it is only the green parts of the plant which, under the influence of sunlight, have the marvellous power of decomposing carbonic acid, setting free the oxygen and laying hold of the carbon which it contains. In fact, the bean obtains two of the absolutely indispensable elements of its substance from two distinct sources; the watery solution, in which its roots are plunged, contains nitrogen but no carbon; the air, to which the leaves are exposed, contains carbon, but its nitrogen is in the state of a free gas, in which condition the bean can make no use of it;[1] and the chlorophyll[2] is the apparatus by which the carbon is extracted from the atmospheric carbonic acid—the leaves being the chief laboratories in which this operation is effected.

The great majority of conspicuous plants are, as everybody knows, green; and this arises from the

[1] I purposely assume that the air with which the bean is supplied in the case stated contains no ammoniacal salts.

[2] The recent researches of Pringsheim have raised a host of questions as to the exact share taken by chlorophyll in the chemical operations which are effected by the green parts of plants. It may be that the chlorophyll is only a constant concomitant of the actual deoxidising apparatus.

abundance of their chlorophyll. The few which contain no chlorophyll and are colourless, are unable to extract the carbon which they require from atmospheric carbonic acid, and lead a parasitic existence upon other plants; but it by no means follows, often as the statement has been repeated, that the manufacturing power of plants depends on their chlorophyll, and its interaction with the rays of the sun. On the contrary, it is easily demonstrated, as Pasteur first proved, that the lowest fungi, devoid of chlorophyll, or of any substitute for it, as they are, nevertheless possess the characteristic manufacturing powers of plants in a very high degree. Only it is necessary that they should be supplied with a different kind of raw material; as they cannot extract carbon from carbonic acid, they must be furnished with something else that contains carbon. Tartaric acid is such a substance; and if a single spore of the commonest and most troublesome of moulds—*Penicillium*—be sown in a saucerful of water, in which tartrate of ammonia, with a small percentage of phosphates and sulphates is contained, and kept warm, whether in the dark or exposed to light, it will, in a short time, give rise to a thick crust of mould, which contains many million times the weight of the original spore, in protein compounds and cellulose. Thus we have a very wide basis of fact for the generalisation that plants are essentially characterised by their manufacturing capacity—by their power of working up mere mineral matters into complex organic compounds.

Contrariwise, there is a no less wide foundation

for the generalisation that animals, as Cuvier puts it, depend directly or indirectly upon plants for the materials of their bodies; that is, either they are herbivorous, or they eat other animals which are herbivorous.

But for what constituents of their bodies are animals thus dependent upon plants? Certainly not for their horny matter; nor for chondrin, the proximate chemical element of cartilage; nor for gelatine; nor for syntonin, the constituent of muscle; nor for their nervous or biliary substances; nor for their amyloid matters; nor, necessarily, for their fats.

It can be experimentally demonstrated that animals can make these for themselves. But that which they cannot make, but must, in all known cases, obtain directly or indirectly from plants, is the peculiar nitrogenous matter, protein. Thus the plant is the ideal *prolétaire* of the living world, the worker who produces; the animal, the ideal aristocrat, who mostly occupies himself in consuming, after the manner of that noble representative of the line of Zähdarm, whose epitaph is written in *Sartor Resartus.*

Here is our last hope of finding a sharp line of demarcation between plants and animals; for, as I have already hinted, there is a border territory between the two kingdoms, a sort of no-man's-land, the inhabitants of which certainly cannot be discriminated and brought to their proper allegiance in any other way.

Some months ago, Professor Tyndall asked me to examine a drop of infusion of hay, placed under an excellent and powerful microscope, and to tell him what I thought some organisms visible in it were. I looked and observed, in the first place, multitudes of *Bacteria* moving about with their ordinary intermittent spasmodic wriggles. As to the vegetable nature of these there is now no doubt. Not only does the close resemblance of the *Bacteria* to unquestionable plants, such as the *Oscillatoriæ*, and lower forms of *Fungi*, justify this conclusion, but the manufacturing test settles the question at once. It is only needful to add a minute drop of fluid containing *Bacteria*, to water in which tartrate, phosphate, and sulphate of ammonia are dissolved; and, in a very short space of time, the clear fluid becomes milky by reason of their prodigious multiplication, which, of course, implies the manufacture of living Bacterium-stuff out of these merely saline matters.

But other active organisms, very much larger than the *Bacteria*, attaining in fact the comparatively gigantic dimensions of $\frac{1}{3000}$ of an inch or more, incessantly crossed the field of view. Each of these had a body shaped like a pear, the small end being slightly incurved and produced into a long curved filament, or *cilium*, of extreme tenuity. Behind this, from the concave side of the incurvation, proceeded another long cilium, so delicate as to be discernible only by the use of the highest powers and careful management of the light. In the centre of the pear-shaped body a clear round space could occasionally be discerned, but not always; and careful watching showed that this clear vacuity appeared gradually, and then shut up and disappeared suddenly, at regular intervals. Such a structure is of common occurrence among the lowest plants and animals, and is known as a *contractile vacuole.*

The little creature thus described sometimes propelled itself with great activity, with a curious rolling motion, by the lashing of the front cilium, while the second cilium trailed behind; sometimes it anchored itself by the hinder cilium and was spun round by the working of the other, its motions resembling those of an anchor buoy in a heavy sea. Sometimes, when two were in full career towards one another, each would appear dexterously to get out of the other's way; sometimes a crowd would assemble and jostle one another, with as much semblance of individual effort as a spectator on the Grands Mulets might observe with a telescope among the specks representing men in the valley of Chamounix.

The spectacle, though always surprising, was not new to me. So my reply to the question put to me was, that these organisms were what biologists call *Monads,* and though they might be animals, it was also possible that they might, like the *Bacteria,* be plants. My friend received my verdict with an expression which showed a sad want of respect for authority. He would as soon believe that a sheep was a plant. Naturally piqued by this want of faith, I have thought a good deal over the matter; and as I still rest in the lame conclusion I originally expressed, and must even now confess that I cannot certainly say whether this creature is an animal or a plant, I think it may be well to state the grounds of my hesitation at length. But, in the first place, in order that I may conveniently distinguish this "Monad" from the multitude of other things which go by the same designation, I must give it a name of its own. I think (though, for reasons which need not be stated at present, I am not quite sure) that it is identical with the species *Monas lens,* as defined by the eminent French microscopist Dujardin, though his magnifying power was probably insufficient to enable him to see that it is curiously like a much larger form of monad which he has named *Heteromita.* I shall, therefore, call it not *Monas,* but *Heteromita lens.*

I have been unable to devote to my *Heteromita* the prolonged study needful to work out its whole history, which would involve weeks, or it may be months, of unremitting attention. But I the less regret this circumstance, as some remarkable observations recently published by Messrs. Dallinger and Drysdale[1] on certain Monads, relate, in part, to a form so similar to my *Heteromita lens,* that the history of the one may be used to illustrate that of the other. These most patient and painstaking observers, who employed the highest attainable powers of the microscope and, relieving one another, kept watch day and night over the same individual monads, have been enabled to trace out the whole history of their *Heteromita;* which they found in infusions of the heads of fishes of the Cod tribe.

[1] "Researches in the Life-history of a Cercomonad : a Lesson in Biogenesis ;" and "Further Researches in the Life-history of the Monads."—"Monthly Microscopical Journal," 1873.

Of the four monads described and figured by these investigators, one, as I have said, very closely resembles *Heteromita lens* in every particular, except that it has a separately distinguishable central particle or "nucleus," which is not certainly to be made out in *Heteromita lens;* and that nothing is said by Messrs. Dallinger and Drysdale of the existence of a contractile vacuole in this monad, though they describe it in another.

Their *Heteromita,* however, multiplied rapidly by fission. Sometimes a transverse constriction appeared; the hinder half developed a new cilium, and the hinder cilium gradually split from its base to its free end, until it was divided into two; a process which, considering the fact that this fine filament cannot be much more than $\frac{1}{100000}$ of an inch in diameter, is wonderful enough. The constriction of the body extended inwards until the two portions were united by a narrow isthmus; finally, they separated and each swam away by itself, a complete *Heteromita,* provided with its two cilia. Sometimes the constriction took a longitudinal direction, with the same ultimate result. In each case the process occupied not more than six or seven minutes. At this rate, a single *Heteromita* would give rise to a thousand like itself in the course of an hour, to about a million in two hours, and to a number greater than the generally assumed number of human beings now living in the world in three hours; or, if we give each *Heteromita* an hour's enjoyment of individual existence, the same result will be obtained in about a day.

The apparent suddenness of the appearance of multitudes of such organisms as these, in any nutritive fluid to which one obtains access, is thus easily explained.

During these processes of multiplication by fission, the *Heteromita* remains active; but sometimes another mode of fission occurs. The body becomes rounded and quiescent, or nearly so ; and, while in this resting state, divides into two portions, each of which is rapidly converted into an active *Heteromita*.

A still more remarkable phenomenon is that kind of multiplication which is preceded by the union of two monads, by a process which is termed *conjugation*. Two active *Heteromitæ* become applied to one another, and then slowly and gradually coalesce into one body. The two nuclei run into one ; and the mass resulting from the conjugation of the two *Heteromitæ*, thus fused together, has a triangular form. The two pairs of cilia are to be seen, for some time, at two of the angles, which answer to the small ends of the conjoined monads ; but they ultimately vanish, and the twin organism, in which all visible traces of organisation have disappeared, falls into a state of rest. Sudden wave-like movements of its substance next occur ; and, in a short time, the apices of the triangular mass burst, and give exit to a dense yellowish, glairy fluid, filled with minute granules. This process, which, it will be observed, involves the actual confluence and mixture of the substance of two distinct organisms, is effected in the space of about two hours.

The authors whom I quote say that they "cannot

express" the excessive minuteness of the granules in question, and they estimate their diameter at less than $\frac{1}{200000}$ of an inch. Under the highest powers of the microscope at present applicable such specks are hardly discernible. Nevertheless, particles of this size are massive when compared to physical molecules; whence there is no reason to doubt that each, small as it is, may have a molecular structure sufficiently complex to give rise to the phenomena of life. And, as a matter of fact, by patient watching of the place at which these infinitesimal living particles were discharged, our observers assured themselves of their growth and development into new monads. These, in about four hours from their being set free, had attained a sixth of the length of the parent, with the characteristic cilia, though at first they were quite motionless; and, in four hours more, they had attained the dimensions and exhibited all the activity of the adult. These inconceivably minute particles are therefore the germs of the *Heteromita;* and from the dimensions of these germs it is easily shown that the body formed by conjugation may, at a low estimate, have given exit to thirty thousand of them; a result of a matrimonial process whereby the contracting parties, without a metaphor, "become one flesh," enough to make a Malthusian despair of the future of the Universe.

I am not aware that the investigators from whom I have borrowed this history have endeavoured to ascertain whether their monads take solid nutriment or not; so that though they help us very much to fill up the blanks in the history of my *Heteromita*, their observations throw no light on the problem we are trying to solve—Is it an animal or is it a plant?

Undoubtedly it is possible to bring forward very strong arguments in favour of regarding *Heteromita* as a plant.

For example, there is a Fungus, an obscure and almost microscopic mould, termed *Peronospora infestans*. Like many other Fungi, the *Peronosporæ* are parasitic upon other plants; and this particular *Peronospora* happens to have attained much notoriety and political importance, in a way not without a parallel in the career of notorious politicians, namely, by reason of the frightful mischief it has done to mankind. For it is this *Fungus* which is the cause of the potato disease; and, therefore, *Peronospora infestans* (doubtless of exclusively Saxon origin, though not accurately known to be so) brought about the Irish famine. The plants afflicted with the malady are found to be infested by a mould, consisting of fine tubular filaments, termed *hyphæ*, which burrow through the substance of the potato plant, and appropriate to themselves the substance of their host; while, at the same time, directly or indirectly, they set up chemical changes by which even its woody framework becomes blackened, sodden, and withered.

In structure, however, the *Peronospora* is as much a mould as the common *Penicillium;* and just as the *Penicillium* multiplies by the breaking up of its hyphæ into separate rounded bodies, the spores; so, in the

Peronospora, certain of the hyphæ grow out into the air through the interstices of the superficial cells of the potato plant, and develop spores. Each of these hyphæ usually gives off several branches. The ends of the branches dilate and become closed sacs, which eventually drop off as spores. The spores falling on some part of the same potato plant, or carried by the wind to another, may at once germinate, throwing out tubular prolongations which become hyphæ, and burrow into the substance of the plant attacked. But, more commonly, the contents of the spore divide into six or eight separate portions. The coat of the spore gives way, and each portion then emerges as an independent organism, which has the shape of a bean, rather narrower at one end than the other, convex on one side, and depressed or concave on the opposite. From the depression, two long and delicate cilia proceed, one shorter than the other, and directed forwards. Close to the origin of these cilia, in the substance of the body, is a regularly pulsating, contractile vacuole. The shorter cilium vibrates actively, and effects the locomotion of the organism, while the other trails behind; the whole body rolling on its axis with its pointed end forwards.

The eminent botanist, De Bary, who was not thinking of our problem, tells us, in describing the movements of these "Zoospores," that, as they swim about, "Foreign bodies are carefully avoided, and the whole movement has a deceptive likeness to the voluntary changes of place which are observed in microscopic animals."

After swarming about in this way in the moisture on the surface of a leaf or stem (which, film though it may be, is an ocean to such a fish) for half an hour, more or less, the movement of the zoospore becomes slower, and is limited to a slow turning upon its axis, without change of place. It then becomes quite quiet, the cilia disappear, it assumes a spherical form, and surrounds itself with a distinct, though delicate, membranous coat. A protuberance then grows out from one side of the sphere, and rapidly increasing in length, assumes the character of a hypha. The latter penetrates into the substance of the potato plant, either by entering a stomate, or by boring through the wall of an epidermic cell, and ramifies, as a mycelium, in the substance of the plant, destroying the tissues with which it comes in contact. As these processes of multiplication take place very rapidly, millions of spores are soon set free from a single infested plant; and, from their minuteness, they are readily transported by the gentlest breeze. Since, again, the zoospores set free from each spore, in virtue of their powers of locomotion, swiftly disperse themselves over the surface, it is no wonder that the infection, once started, soon spreads from field to field, and extends its ravages over a whole country.

However, it does not enter into my present plan to treat of the potato disease, instructively as its history bears upon that of other epidemics; and I have selected the case of the *Peronospora* simply because it affords an example of an organism, which, in one stage of its existence, is truly a "Monad," indistinguishable by any important character from our *Heteromita*, and extraordinarily like it in some respects. And yet this "Monad" can be traced, step by step, through the series of metamorphoses which I have described, until it assumes the features of an organism, which is as much a plant as is an oak or an elm.

Moreover, it would be possible to pursue the analogy farther. Under certain circumstances, a process of conjugation takes place in the *Peronospora*. Two separate portions of its protoplasm become fused together, surround themselves with a thick coat, and give rise to a sort of vegetable egg called an *oospore*. After a period of rest, the contents of the oospore break up into a number of zoospores like those already described, each of which, after a period of activity, germinates in the ordinary way. This process obviously corresponds with the conjugation and subsequent setting free of germs in the *Heteromita*.

But it may be said that the *Peronospora* is, after all, a questionable sort of plant; that it seems to be wanting in the manufacturing power, selected as the main distinctive character of vegetable life; or, at any rate, that there is no proof that it does not get its protein matter ready made from the potato plant.

Let us, therefore, take a case which is not open to these objections.

There are some small plants known to botanists as members of the genus *Coleochæte*, which, without being truly parasitic, grow upon certain water-weeds, as lichens grow upon trees. The little plant has the form of an elegant green star, the branching arms of which are divided into cells. Its greenness is due to its chlorophyll, and it undoubtedly has the manufacturing power in full degree, decomposing carbonic acid and setting oxygen free, under the influence of sunlight. But the protoplasmic contents of some of the cells of which the plant is made up occasionally divide, by a method similar to that which effects the division of the contents of the *Peronospora* spore; and the severed portions are then set free as active monad-like zoospores. Each is oval and is provided at one extremity with two long active cilia. Propelled by these, it swims about for a longer or shorter time, but at length comes to a state of rest and gradually grows into a *Coleochæte*. Moreover, as in the *Peronospora*, conjugation may take place and result in an oospore; the contents of which divide and are set free as monadiform germs.

If the whole history of the zoospores of *Peronospora* and of *Coleochæte* were unknown, they would undoubtedly be classed among "Monads" with the same right as *Heteromita*; why then may not *Heteromita* be a plant, even though the cycle of forms through which it passes shows no terms quite so complex as those which occur in *Peronospora* and *Coleochæte*? And, in fact, there are some green organisms, in every respect characteristically plants, such as *Chlamydomonas*, and the common *Volvox*, or so-called "Globe animalcule," which run through a cycle of forms of just the same simple character as those of *Heteromita*.

The name of *Chlamydomonas* is applied to certain

microscopic green bodies, each of which consists of a protoplasmic central substance invested by a structureless sac. The latter contains cellulose, as in ordinary plants; and the chlorophyll which gives the green colour enables the *Chlamydomonas* to decompose carbonic acid and fix carbon as they do. Two long cilia protrude through the cell-wall, and effect the rapid locomotion of this "monad," which, in all respects except its mobility, is characteristically a plant. Under ordinary circumstances, the *Chlamydomonas* multiplies by simple fission, each splitting into two or into four parts, which separate and become independent organisms. Sometimes, however, the *Chlamydomonas* divides into eight parts, each of which is provided with four instead of two cilia. These "zoospores" conjugate in pairs, and give rise to quiescent bodies, which multiply by division, and eventually pass into the active state.

Thus, so far as outward form and the general character of the cycle of modifications, through which the organism passes in the course of its life, are concerned, the resemblance between *Chlamydomonas* and *Heteromita* is of the closest description. And on the face of the matter there is no ground for refusing to admit that *Heteromita* may be related to *Chlamydomonas*, as the colourless fungus is to the green alga. *Volvox* may be compared to a hollow sphere, the wall of which is made up of coherent Chlamydomonads; and which progresses with a rotating motion effected by the paddling of the multitudinous pairs of cilia which project from its surface. Each *Volvox*-

monad, moreover, possesses a red pigment spot, like the simplest form of eye known among animals. The methods of fissive multiplication and of conjugation observed in the monads of this locomotive globe are essentially similar to those observed in *Chlamydomonas*; and, though a hard battle has been fought over it, *Volvox* is now finally surrendered to the Botanists.

Thus there is really no reason why *Heteromita* may not be a plant; and this conclusion would be very satisfactory, if it were not equally easy to show that there is really no reason why it should not be an animal. For there are numerous organisms presenting the closest resemblance to *Heteromita*, and, like it, grouped under the general name of "Monads," which, nevertheless, can be observed to take in solid nutriment, and which, therefore, have a virtual, if not an actual, mouth and digestive cavity, and thus come under Cuvier's definition of an animal. Numerous forms of such animals have been described by Ehrenberg, Dujardin, H. James Clark, and other writers on the *Infusoria*. Indeed, in another infusion of hay in which my *Heteromita lens* occurred, there were innumerable infusorial animalcules belonging to the well-known species *Colpoda cucullus*.[1]

Full-sized specimens of this animalcule attain a length of between $\frac{1}{300}$ or $\frac{1}{400}$ of an inch, so that it may have ten times the length and a thousand times the mass of a *Heteromita*. In shape, it is not alto-

[1] Excellently described by Stein, almost all of whose statements I have verified.

gether unlike *Heteromita.* The small end, however, is not produced into one long cilium, but the general surface of the body is covered with small actively vibrating ciliary organs, which are only longest at the small end. At the point which answers to that from which the two cilia arise in *Heteromita,* there is a conical depression, the mouth; and, in young specimens, a tapering filament, which reminds one of the posterior cilium of *Heteromita,* projects from this region.

The body consists of a soft granular protoplasmic substance, the middle of which is occupied by a large oval mass called the "nucleus;" while, at its hinder end, is a "contractile vacuole," conspicuous by its regular rhythmic appearances and disappearances. Obviously, although the *Colpoda* is not a monad, it differs from one only in subordinate details. Moreover, under certain conditions, it becomes quiescent, incloses itself in a delicate case or *cyst,* and then divides into two, four, or more portions, which are eventually set free and swim about as active *Colpodæ.*

But this creature is an unmistakable animal, and full-sized *Colpodæ* may be fed as easily as one feeds chickens. It is only needful to diffuse very finely ground carmine through the water in which they live, and, in a very short time, the bodies of the *Colpodæ* are stuffed with the deeply-coloured granules of the pigment.

And if this were not sufficient evidence of the animality of *Colpoda,* there comes the fact that it is even more similar to another well-known animalcule,

Paramœcium, than it is to a monad. But *Paramœcium* is so huge a creature compared with those hitherto discussed—it reaches $\frac{1}{120}$ of an inch or more in length—that there is no difficulty in making out its organisation in detail; and in proving that it is not only an animal, but that it is an animal which possesses a somewhat complicated organisation. For example, the surface layer of its body is different in structure from the deeper parts. There are two contractile vacuoles, from each of which radiates a system of vessel-like canals; and not only is there a conical depression continuous with a tube, which serve as mouth and gullet, but the food ingested takes a definite course, and refuse is rejected from a definite region. Nothing is easier than to feed these animals, and to watch the particles of indigo or carmine accumulate at the lower end of the gullet. From this they gradually project, surrounded by a ball of water, which at length passes with a jerk, oddly simulating a gulp, into the pulpy central substance of the body, there to circulate up one side and down the other, until its contents are digested and assimilated. Nevertheless, this complex animal multiplies by division, as the monad does, and, like the monad, undergoes conjugation. It stands in the same relation to *Heteromita* on the animal side, as *Coleochæte* does on the plant side. Start from either, and such an insensible series of gradations leads to the monad that it is impossible to say at any stage of the progress—here the line between the animal and the plant must be drawn.

There is reason to think that certain organisms which pass through a monad stage of existence, such as the *Myxomycetes*, are, at one time of their lives, dependent upon external sources for their protein matter, or are animals; and, at another period, manufacture it, or are plants. And seeing that the whole progress of modern investigation is in favour of the doctrine of continuity, it is a fair and probable speculation—though only a speculation—that, as there are some plants which can manufacture protein out of such apparently intractable mineral matters as carbonic acid, water, nitrate of ammonia, metallic and earthy salts; while others need to be supplied with their carbon and nitrogen in the somewhat less raw form of tartrate of ammonia and allied compounds; so there may be yet others, as is possibly the case with the true parasitic plants, which can only manage to put together materials still better prepared—still more nearly approximated to protein—until we arrive at such organisms as the *Psorospermiæ* and the *Panhistophyton*, which are as much animal as vegetable in structure, but are animal in their dependence on other organisms for their food.

The singular circumstance observed by Meyer, that the *Torula* of yeast, though an indubitable plant, still flourishes most vigorously when supplied with the complex nitrogenous substance, pepsin; the probability that the *Peronospora* is nourished directly by the protoplasm of the potato-plant; and the wonderful facts which have recently been brought to light respecting insectivorous plants, all favour this view;

and tend to the conclusion that the difference between animal and plant is one of degree rather than of kind; and that the problem whether, in a given case, an organism is an animal or a plant, may be essentially insoluble.

VIII.

ON CERTAIN ERRORS RESPECTING THE STRUCTURE OF THE HEART ATTRIBUTED TO ARISTOTLE.

IN all the commentaries upon the "Historia Animalium" which I have met with, Aristotle's express and repeated statement, that the heart of man and the largest animals contains only three cavities, is noted as a remarkable error. Even Cuvier, who had a great advantage over most of the commentators in his familiarity with the subject of Aristotle's description, and whose habitual caution and moderation seem to desert him when the opportunity of panegyrising the philosopher presents itself, is betrayed into something like a sneer on this topic.

"Du reste il n'attribue à cet organe que trois cavités, erreur qui prouve au moins qu'il en avait regardé la structure."[1]

To which remark, what follows will, I think, justify the reply, that it "prouve au moins" that Cuvier had not given ordinary attention, to say nothing of the careful study which they deserve, to sundry passages in the first and the third books of the "Historia" which I proceed to lay before the reader.

[1] "Histoire des Sciences Naturelles," i. p. 152.

For convenience of reference these passages are marked *A*, *B*, *C*, etc.[1]

Book i. 17.—(*A*) "The heart has three cavities, it lies above the lung on the division of the windpipe, and has a fatty and thick membrane where it is united with the great vein and the aorta. It lies upon the aorta, with its point down the chest, in all animals that have a chest. In all, alike in those that have a chest and in those that have none, the foremost part of it is the apex. This is often overlooked through the turning upside down of the dissection. The rounded end of the heart is uppermost, the pointed end of it is largely fleshy and thick, and in its cavities there are tendons. In other animals which have a chest the heart lies in the middle of the chest; in men, more to the left side, between the nipples, a little inclined to the left nipple in the upper part of the chest. The heart is not large, and its general form is not elongated but rounded, except that the apex is produced into a point.

(*B*) "It has, as already stated, three cavities, the largest of them is on the right, the smallest on the left, the middle-sized one in the middle; they have all, also the two small ones, passages (τετρημένας) towards the lung, very evidently as respects one of the cavities. In the region of the union [with the great vein and the aorta] the largest cavity is connected with the largest vein (near which is the mesentery); the middle cavity with the aorta.

(*C*) "Canals (πόροι) from the heart pass to the lung and divide in the same fashion as the windpipe does, closely accompanying those from the windpipe through the whole lung. The canals from the heart are uppermost.

(*D*) "No canal is common [to the branches of the windpipe and those of the vein] (οὐδεὶς δ' ἐστὶ κοινὸς πόρος) but through

---

[1] The text I have followed is that given by Aubert and Wimmer, "Aristoteles Thierkunde; kritisch berichtigter Text mit deutschen Uebersetzung;" but I have tried here and there to bring the English version rather closer to the original than the German translation, excellent as it is, seems to me to be.

those parts of them which are in contact (τὴν σύναψιν) the air passes in and they [the πόροι] carry it to the heart.

(*E*) "One of the canals leads to the right cavity, the other to the left."

(*F*) "Of all the viscera, the heart alone contains blood [in itself]. The lung contains blood, not in itself but in the veins, the heart in itself; for in each of the cavities there is blood; the thinnest is in the middle cavity."

Book iii. 3.—(*G*) "Two veins lie in the thorax alongside the spine, on its inner face; the larger more forwards, the smaller behind; the larger more to the right, the smaller, which some call *aorta* (on account of the tendinous part of it seen in dead bodies), to the left. These take their origin from the heart; they pass entire, preserving the nature of veins, through the other viscera that they reach; while the heart is rather a part of them, and more especially of the anterior and larger one, which is continued into veins above and below, while between these is the heart.

(*H*) "All hearts contain cavities, but, in those of very small animals, the largest [cavity] is hardly visible, those of middling size have another, and the biggest all three.

(*I*) "The point of the heart is directed forwards, as was mentioned at first; the largest cavity to the right and upper side of it, the smallest to the left, and the middle-sized one between these; both of these are much smaller than the largest.

(*K*) "They are all connected by passages (συντέτρηνται) with the lung, but, on account of the smallness of the canals, this is obscure except in one.

(*L*) "The great vein proceeds from the largest cavity which lies upwards and to the right; next through the hollow middle part (διὰ τοῦ κοῖλου τοῦ μέσου) it becomes vein again, this cavity being a part of the vein in which the blood stagnates.

(*M*) "The aorta [proceeds from] the middle [cavity], but not in the same way, for it is connected [with the middle cavity] by a much more narrow tube (σύριγγα).

(*N*) "The [great] vein extends through the heart, towards the aorta from the heart.

(*O*) "The great vein is membranous like skin, the aorta narrower than it and very tendinous, and as it extends towards the head and the lower parts it becomes narrow and altogether tendinous.

(*P*) "In the first place, a part of the great vein extends upwards from the heart towards the lung and the attachment of the aorta, the vein being large and undivided. It divides into two parts, the one to the lung, the other to the spine and the lowest vertebra of the neck.

(*Q*) "The vein which extends to the lung first divides into two parts for the two halves of it and then extends alongside each tube (σύριγγα) and each passage (τρῆμα), the larger beside the larger and the smaller beside the smaller, so that no part [of the lung] can be found from which a passage (τρῆμα) and a vein are absent. The terminations are invisible on account of their minuteness, but the whole lung appears full of blood. The canals from the vein lie above the tubes given off from the windpipe."

The key to the whole of the foregoing description of the heart lies in the passages (*G*) and (*L*). They prove that Aristotle, like Galen, five hundred years afterwards, and like the great majority of the old Greek anatomists, did not reckon what we call the right auricle as a constituent of the heart at all, but as a hollow part, or dilatation, of the "great vein." Aristotle is careful to state that his observations were conducted on suffocated animals; and if any one will lay open the thorax of a dog or a rabbit, which has been killed with chloroform, in such a manner as to avoid wounding any important vessel, he will at once see why Aristotle adopted this view.

For, as the subjoined figure (p. 185) shows, the vena cava inferior (*b*), the right auricle (*R.a.*), and the vena cava superior and innominate vein (*V.I.*) distended with blood seem to form one continuous column, to which the heart is attached as a sort of appendage. This column is, as Aristotle says, vein above (*a*) and vein below (*b*), the upper and the lower divisions being connected *διὰ τοῦ κοῖλου τοῦ μέσου*—or by means of the intervening cavity or chamber (*R.a.*)—which is that which we call the right auricle.

But when, from the four cavities of the heart recognised by us moderns, one is excluded, there remain three—which is just what Aristotle says. The solution of the difficulty is, in fact, as absurdly simple as that presented by the egg of Columbus; and any error there may be, is not to be put down to Aristotle, but to that inability to comprehend that the same fact may be accurately described in different ways, which is the special characteristic of the commentatorial mind. That the three cavities mentioned by Aristotle are just those which remain if the right auricle is omitted, is plain enough from what is said in (*B*), (*C*), (*E*), (*I*), and (*L*). For, in a suffocated animal, the "right cavity" which is directly connected with the great vein, and is obviously the right ventricle, being distended with blood, will look much larger than the middle cavity, which, since it gives rise to the aorta, can only be the left ventricle. And this, again, will appear larger than the thin and collapsed left auricle, which must be Aristotle's left cavity, inasmuch as this cavity is said to be connected by *πόροι* with the lung. The reason why Aristotle considered the left auricle to be a part of the heart, while he merged the right auricle in the great vein, is, obviously, the small relative size of the

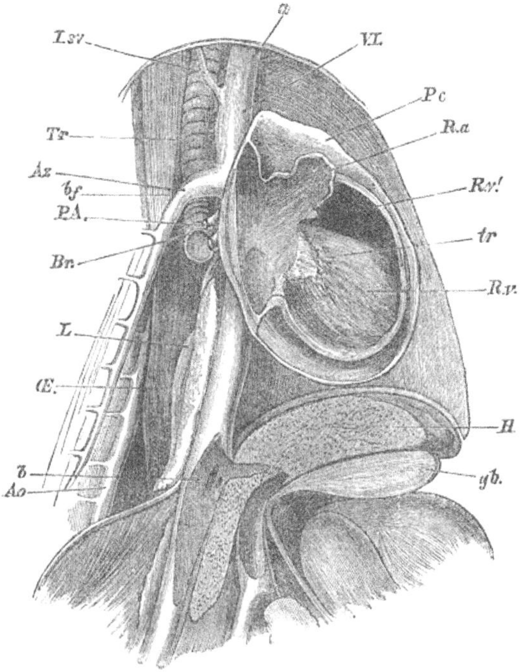

A dog having been killed by chloroform, enough of the right wall of the thorax was removed, without any notable bleeding, to expose the thoracic viscera. A carefully measured outline sketch of the parts *in situ* was then made, and on dissection, twenty-four hours afterwards, the necessary anatomical details were added. The woodcut is a faithfully reduced copy of the drawing thus constructed; and it represents the relations of the heart and great vessels as Aristotle saw them in a suffocated animal.

All but the inner lobe of the right lung has been removed; as well as the right half of the pericardium and the right walls of the right auricle and ventricle. It must be remembered that the thin transparent pericardial membrane appears nothing like so distinct in nature.

a.b., Aristotle's "great vein"; *V.I.*, right vena innominata and vena cava superior; *b*, the inferior vena cava; *R.a.*, the "hollow middle" part of the great vein or the right auricle; *R.v'*, the prolongation of the cavity of the right ventricle *R.v* towards the pulmonary artery; *tr*, one of the tricuspid valves; *Pc*, the pericardium; *I.sv*, superior intercostal vein; *Az*, vena azygos; *P.A.*, right pulmonary artery; *Br*, right bronchus; *L*, inner lobe of the right lung; *Œ*, œsophagus; *Ao*, descending aorta; *H*, liver, in section, with hepatic vein, vena portæ, and gall-bladder, *gb*, separated by the diaphragm, also seen in section, from the thoracic cavity.

venous trunks and their sharper demarcation from the auricle. Galen, however, perhaps more consistently, regarded the left auricle also as a mere part of the "arteria venosa." The canal which leads from the right cavity of the heart to the lung (or, as Aristotle puts it (*E*), from the lung to the heart) is, without doubt, the pulmonary artery. But it may be said that, in this case, Aristotle contradicts himself, inasmuch as in (*P*) and (*Q*) a vessel, which is obviously the pulmonary artery, is described as a branch of the great vein. However, this difficulty also disappears, if we reflect that, in Aristotle's way of looking at the matter, the line of demarcation between the great vein and the heart coincides with the right auriculo-ventricular aperture; and that, inasmuch as the conical prolongation of the right ventricle which leads to the pulmonary artery (*R.v′* in the Figure), lies close in front of the auricle, its base may very easily (as the figure shows) be regarded as part of the general opening of the great vein into the right ventricle. In fact, it is clear that Aristotle, having failed to notice the valves of the heart, did not distinguish the part of the right ventricle from which the pulmonary artery arises (*R.v′*) from the proper trunk of the artery on the one hand, and from the right auricle (*R.a*) on the other. Thus the root, as we may call it, of the pulmonary artery and the right auricle, taken together, are spoken of as the "part of the great vein which extends upwards" (*P*); and, as the vena azygos (*Az*) was one branch of this, so the "vein to the lung" was regarded as another branch of it. But the latter branch, being given off close to the connection of the great vein with the ventricle, was also counted as one of the two *πόροι* by which the "heart" (that is to say the right ventricle, the left ventricle, and the left auricle of our nomenclature) communicates with the lung.

The only other difficulty that I observe is connected with (*K*). If Aristotle intended by this to affirm that the middle cavity (the left ventricle), like the other two, is directly connected with the lung by a *πόρος*, he would be in error. But he has excluded this interpretation of his words by (*E*), in which the number and relations of the canals, the existence of which he admits, are distinctly defined. I can only imagine then, that, so far as this passage applies to the left ventricle, it merely refers to the indirect communication of that cavity with the vessels of the lungs, through the left auricle.

On this evidence I submit that there is no escape from the conclusion that, instead of having committed a gross blunder, Aristotle has given a description of the heart which, so far as it goes, is remarkably accurate. He is in error only in regard to the differences which he imagines to exist between large and small hearts (*H*).

Cuvier (who has been followed by other commentators) ascribes another error to Aristotle :—

> "Aristote suppose que la trachée-artère se prolonge jusqu'au cœur, et semble croire, en conséquence, que l'air y pénétre (*l. c.* p. 152)."

Upon what foundation Cuvier rested the first of

these two assertions, I am at a loss to divine. As a matter of fact, it will appear from the following excerpts that Aristotle gives an account of the structure of the lungs which is almost as good as that of the heart, and that it contains nothing about any prolongation of the windpipe to the heart.

"Within the neck lie what is called the œsophagus (so named on account of its length and its narrowness) and the windpipe (ἀρτηρία). The position of the windpipe in all animals that have one, is in front of the œsophagus. All animals which possess a lung have a windpipe. The windpipe is of a cartilaginous nature and is exsanguine, but is surrounded by many little veins. . . .

"It goes downwards towards the middle of the lung, and then divides for each of the halves of the lung. In all animals that possess one, the lung is divided into two parts; but, in those which bring forth their young alive, the separation is not equally well marked, least of all in man.

"In oviparous animals, such as birds, and in quadrupeds which are oviparous, the one half of the lung is widely separated from the other; so that it appears as if they had two lungs. And from being single, the windpipe becomes (divided into) two, which extend to each half of the lung. It is fastened to the great vein, and to what is called the aorta. When the windpipe is blown up, the air passes into the hollow parts of the lung. In these, are cartilaginous tubes (διαφύσεις) which unite at an angle; from the tubes passages (τρήματα) traverse the whole of the lung; they are continually given off, the smaller from the larger." (Book i. 16.)

That Aristotle should speak of the lung as a single organ divided into two halves, and should say that the division is least marked in man, is puzzling at first; but the statement becomes intelligible, if we reflect upon the close union of the bronchi, the pulmonary vessels and the mediastinal walls of the pleuræ, in mammals;[1] and it is quite true that the lungs are much more obviously distinct from one another in birds.

Aubert and Wimmer translate the last paragraph of the passage just cited as follows:—

"Diese haben aber knorpelige Scheidewände, welche unter spitzen Winkeln zusammentreten, und aus ihnen führen Oeffnungen durch die ganze Lunge, indem sie sich in immer kleineren verzweigen."

But I cannot think that by *διαφύσεις* and *τρήματα*, in this passage, Aristotle meant either "partitions" or openings in the ordinary sense of the latter word. For, in Book iii. Cap. 3, in describing the distribution of the "vein which goes to the lung" (the pulmonary artery), he says that it

"extends alongside each tube (*σύριγγα*) and each passage (*τρῆμα*), the larger beside the larger, and the smaller beside the smaller; so that no part (of the lung) can be found from which a passage (*τρῆμα*) and a vein are absent."

Moreover, in Book i. 17, he says—

"Canals (*πόροι*) from the heart pass to the lung and divide in the same fashion as the windpipe does, closely accompanying those from the windpipe through the whole lung."

And again in Book i. 17—

"It (the lung) is entirely spongy, and alongside of each tube (*σύριγγα*) run canals (*πόροι*) from the great vein."

On comparing the last three statements with the facts of the case, it is plain that by *σύριγγες*, or tubes, Aristotle means the bronchi and so many of their

[1] In modern works on Veterinary Anatomy the lungs are sometimes described as two lobes of a single organ.

larger divisions as obviously contain cartilages; and that by *διαφύσεις χονδρώδεις* he denotes the same things; and, if this be so, then the *τρήματα* must be the smaller bronchial canals, in which the cartilages disappear.

This view of the structure of the lung is perfectly correct so far as it extends; and, bearing it in mind, we shall be in a position to understand what Aristotle thought about the passage of air from the lungs into the heart. In every part of the lung, he says, in effect, there is an air tube which is derived from the trachea, and other tubes which are derived from the *πόροι* which connect the lung with the heart (*suprà*, *C*). Their applied walls constitute the thin "synapses" (*τὴν σύναψιν*) through which the air passes out of the air tubes into the *πόροι*, or blood-vessels, by transudation or diffusion; for there is no community between the cavities of the air tubes and cavities of the canals; that is to say, no opening from one into the other (*suprà*, *D*).

On the words "*κοινὸς πόρος*" Aubert and Wimmer remark (*l. c.* p. 239), "Da A. die Ansicht hat die Lungenluft würde dem Herzen zugeführt, so postulirt er statt vieler kleiner Verbindungen einen grossen Verbindungsgang zwischen Lunge und Herz."

But does Aristotle make this assumption? The only evidence so far as I know in favour of the affirmative answer to this question is the following passage:—

"Συνήρτεται δὲ καὶ ἡ καρδία τῇ ἀρτηρίᾳ πιμελώδεσι καὶ χονδρώδεσι καὶ ἰνώδεσι δεσμοῖς· ᾗ δέ συνήρτεται, κοῖλόν ἐστιν. φυσωμένης δὲ τῆς ἀρτηρίας μὲν ἐνίοις ἐν οὐ κατάδηλον ποιεῖ, ἐν δὲ τοῖς μείζοσι τῶν ζῴων δῆλον ὅτι εἰσέρχεται τὸ πνεῦμα εἰς αὐτὴν" (i. cap. 16).

"The heart and the windpipe are connected by fatty and cartilaginous and fibrous bands; where they are connected it is hollow. Blowing into the windpipe does not show clearly in some animals, but in the larger animals it is clear that the air goes into it."

Aubert and Wimmer give a somewhat different rendering of this passage:—

"Auch das Herz hängt mit der Luftröhre durch fettreiche, knorpelige und faserige Bänder zusammen; und da, wo sie zusammenhängen, ist eine Höhlung. Beim Aufblasen der Lunge wird es bei manchen Thieren nicht wahrnehmbar, bei den grösseren aber ist es offenbar, dass die Luft in das Herz gelangt."

The sense here turns upon the signification which is to be ascribed to εἰς αὐτὴν. But if these words refer to the heart, then Aristotle has distinctly pointed out the road which the air, in his opinion, takes, namely, through the "synapses" (*D*); and there is no reason that I can discover to believe that he "postulated" any other and more direct communication.

With respect to the meaning of κοῖλόν ἐστιν, Aubert and Wimmer observe:—

"Dies scheint wohl die kurze Lungenvene zu sein. Schneider bezieht dies auf die Vorkammern, allein diese werden unten als Höhlen des Herzens beschrieben."

I am disposed to think, on the contrary, that the words refer simply to the cavity of the pericardium. For a part of this cavity (*sinus transversus pericardii*) lies between the aorta, on the one hand, and the pulmonary vessels with the bifurcation of the trachea, on the other hand, and is much more conspicuous in some animals than in man. It is strictly correct, therefore, in Aristotle's words, to say that where the heart and the windpipe are connected "it is hollow." If he had meant to speak of one of the pulmonary veins, or of any of the cavities of the heart, he would have used the terms *πόροι* or *κοιλίας* which he always employs for these parts.

According to Aristotle, then, the air taken into the lungs passes, from the final ramifications of the bronchial tubes into the corresponding branches of the pulmonary blood-vessels, not through openings, but by transudation, or, as we should nowadays say, diffusion, through the thin partitions formed by the applied coats of the two sets of canals. But the "pneuma" which thus reached the interior of the blood-vessels was not, in Aristotle's opinion, exactly the same thing as the air. It was "*ἀήρ πολὺς ῥέων καὶ ἀθρόος*" ("De Mundo," iv. 9)—subtilised and condensed air; and it is hard to make out whether Aristotle considered it to possess the physical properties of an elastic fluid or those of a liquid. As he affirms that all the cavities of the heart contain blood (*F*), it is clear that he did not hold the erroneous view propounded in the next generation by Erasistratus. On the other hand, the fact that he supposes that the spermatic arteries do not contain blood but only an *αἱματῶδης ὑγρόν* ("Hist. Animalium," iii. 1), shows that his notions respecting the contents of the arteries were vague. Nor does he seem to have known that the pulse is characteristic only of the arteries; and as he thought that the arteries end in solid fibrous bands, he naturally could not have entertained the faintest conception of the true motion of the blood. But, without attempting to read into Aristotle modern conceptions which never entered his mind, it is only just to observe that his view of what becomes of the air taken into the lungs is by no means worthy of contempt as a gross error. On the contrary, here, as in the case of his anatomy of the heart, what Aristotle asserts is true as far as it goes. Something does actually pass from the air contained in the lungs through the coats of the vessels into the blood, and thence to the heart; to wit, oxygen. And I think that it speaks very well for ancient Greek science that the investigator of so difficult a physiological problem as that of respiration, should have arrived at a conclusion, the statement of which, after the lapse of more than two thousand years, can be accepted as a thoroughly established scientific truth.

I trust that the case in favour of removing the statements about the heart, from the list of the "errors of Aristotle" is now clear; and that the evidence proves, on the contrary, that they justify us in forming a very favourable estimate of the oldest anatomical investigations among the Greeks of which any sufficient record remains.

But is Aristotle to be credited with the merit of having ascertained so much of the truth? This question will not appear superfluous to those who are acquainted with the extraordinary history of Aristotle's works, or who adopt the conclusion of Aubert and Wimmer, that, of the ten books of the "Historia

Animalium" which have come down to us, three are largely or entirely spurious, and that the others contain many interpolations by later writers.

It so happens, however, that, apart from other reasons, there are satisfactory internal grounds for ascribing the account of the heart to a writer of the time at which Aristotle lived. For, within thirty years of his death, the anatomists of the Alexandrian school had thoroughly investigated the structure and the functions of the valves of the heart. During this time, the manuscripts of Aristotle were in the possession of Theophrastus; and no interpolator of later date would have shown that he was ignorant of the nature and significance of these important structures, by the brief and obscure allusion—"in its cavities there are tendons" (*A*). On the other hand, Polybus, whose account of the vascular system is quoted in the "Historia Animalium" was an elder contemporary of Aristotle. Hence, if any part of the work faithfully represents that which Aristotle taught, we may safely conclude that the description of the heart does so. Having granted this much, however, it is another question, whether Aristotle is to be regarded as the first discoverer of the facts which he has so well stated, or whether he, like other men, was the intellectual child of his time and simply carried on a step or two the work which had been commenced by others.

On the subject of Aristotle's significance as an original worker in biology extraordinarily divergent views have been put forward. If we are to adopt Cuvier's estimate, Aristotle was simply a miracle:—

"Avant Aristote la philosophie, entièrement spéculative, se perdait dans les abstractions dépourvues de fondement; la science n'existait pas. Il semble qu'elle soit sortie toute faite du cerveau d'Aristote comme Minerve, toute armée, du cerveau de Jupiter. Seul, en effet, sans antécédents, sans rien emprunter aux siècles qui l'avaient précédé, puisqu'ils n'avaient rien produit de solide, le disciple de Platon découvrit et démontra plus de vérités, executa plus de travaux scientifiques en un vie de soixante-deux ans, qu'après lui vingt siècles n'en ont pu faire,"[1] etc. etc.

"Aristote est le premier qui ait introduit la méthode de l'induction, de la comparaison des observations pour en faire sortir des idées générales, et celle de l'expérience pour multiplier les faits dont ces idées générales peuvent être déduites."—ii. p. 515.

The late Mr. G. H. Lewes,[2] on the contrary, tells us "on a superficial examination, therefore, he [Aristotle] will seem to have given tolerable descriptions; especially if approached with that disposition to discover marvels which unconsciously determines us in our study of eminent writers. But a more unbiassed and impartial criticism will disclose that he has given no single anatomical description of the least value. All that he knew may have been known, and probably was known, without dissection. . . . I do not assert that he never opened an animal; on the contrary it seems highly probable that he had opened many. . . . He never followed the course of a vessel or a nerve; never laid bare the origin and insertion of a muscle; never discriminated the component parts of organs; never made clear to himself the connection of organs into systems."—(pp. 156-7.)

[1] "Histoire des Sciences Naturelles."—t. i. p. 130.

[2] "Aristotle, a Chapter from the History of Science."

In the face of the description of the heart and lungs, just quoted, I think we may venture to say that no one who has acquired even an elementary practical acquaintance with anatomy, and knows of his own knowledge that which Aristotle describes, will agree with the opinion expressed by Mr. Lewes; and those who turn to the accounts of the structure of the rock lobster and the common lobster, or to that of the Cephalopods and other Mollusks, in the fourth book of the "Historia Animalium" will probably feel inclined to object to it still more strongly.

On the other hand, Cuvier's exaggerated panegyric will as little bear the test of cool discussion. In Greece, the century before Aristotle's birth was a period of great intellectual activity, in the field of physical science no less than elsewhere. The method of induction has never been used to better effect than by Hippocrates; and the labours of such men as Alkmeon, Demokritus, and Polybus, among Aristotle's predecessors; Diokles, and Praxagoras, among his contemporaries, laid a solid foundation for the scientific study of anatomy and development, independently of his labours. Aristotle himself informs us that the dissection of animals was commonly practised; that the aorta had been distinguished from the great vein; and that the connection of both with the heart had been observed by his predecessors. What they thought about the structure of the heart itself or that of the lungs, he does not tell us, and we have no means of knowing. So far from arrogantly suggesting that he owed nothing to his predecessors, Aristotle is careful to refer to their observations, and to explain why, in his judgment, they fell into the errors which he corrects.

Aristotle's knowledge, in fact, appears to have stood in the same relation to that of such men as Polybus and Diogenes of Apollonia, as that of Herophilus and Erasistratus did to his own, so far as the heart is concerned. He carried science a step beyond the point at which he found it; a meritorious, but not a miraculous, achievement. What he did, required the possession of very good powers of observation; if they had been powers of the highest class, he could hardly have left such conspicuous objects as the valves of the heart to be discovered by his successors.

And this leads me to make a final remark upon a singular feature of the "Historia Animalium." As a whole, it is a most notable production, full of accurate information, and of extremely acute generalisations of the observations accumulated by naturalists up to that time. And yet, every here and there, one stumbles upon assertions respecting matters which lie within the scope of the commonest inspection, which are not so much to be called errors, as stupidities. What is to be made of the statement that the sutures of women's skulls are different from those of men; that men and sundry male animals have more teeth than their respective females; that the back of the skull is empty; and so on? It is simply incredible to me, that the Aristotle who wrote the account of the heart, also committed himself to absurdities which can be excused by no theoretical prepossession and which are contradicted by the plainest observation.

What, after all, were the original manuscripts of the "Historia Animalium"? If they were notes of Aristotle's lectures taken by some of his students, any lecturer who has chanced to look through such notes, would find the interspersion of a foundation of general and sometimes minute accuracy, with patches of transcendent blundering, perfectly intelligible. Some competent Greek scholar may perhaps think it worth while to tell us what may be said for or against the hypothesis thus hinted. One obvious difficulty in the way of adopting it is the fact that, in other works, Aristotle refers to the "Historia Animalium" as if it had already been made public by himself.

IX.

ON THE HYPOTHESIS THAT ANIMALS ARE AUTOMATA, AND ITS HISTORY.

THE first half of the seventeenth century is one of the great epochs of biological science. For though suggestions and indications of the conceptions which took definite shape, at that time, are to be met with in works of earlier date, they are little more than the shadows which coming truth casts forward; men's knowledge was neither extensive enough, nor exact enough, to show them the solid body of fact which threw these shadows.

But, in the seventeenth century, the idea that the physical processes of life are capable of being explained in the same way as other physical phenomena, and, therefore, that the living body is a mechanism, was proved to be true for certain classes of vital actions; and, having thus taken firm root in irrefragable fact, this conception has not only successfully repelled every assault which has been made upon it, but has steadily grown in force and extent of application, until it is now the expressed or implied fundamental proposition of the whole doctrine of scientific Physiology.

If we ask to whom mankind are indebted for this great service, the general voice will name William Harvey. For, by his discovery of the circulation of the blood in the higher animals, by his explanation of the nature of the mechanism by which that circulation is effected, and by his no less remarkable, though less known, investigations of the process of development, Harvey solidly laid the foundations of all those physical explanations of the functions of sustentation and reproduction which modern physiologists have achieved.

But the living body is not only sustained and reproduced: it adjusts itself to external and internal changes; it moves and feels. The attempt to reduce the endless complexities of animal motion and feeling to law and order is, at least, as important a part of the task of the physiologist as the elucidation of what are sometimes called the vegetative processes. Harvey did not make this attempt himself; but the influence of his work upon the man who did make it is patent and unquestionable. This man was René Descartes, who, though by many years Harvey's junior, died before him; and yet, in his short span of fifty-four years, took an undisputed place, not only among the chiefs of philosophy, but amongst the greatest and most original of mathematicians; while, in my belief, he is no less certainly entitled to the rank of a great and original physiologist; inasmuch as he did for the physiology of motion and sensation that which Harvey had done for the circulation of the blood, and opened up that road to the mechanical theory of these processes, which has been followed by all his successors.

Descartes was no mere speculator, as some would have us believe: but a man who knew of his own knowledge what was to be known of the facts of anatomy and physiology in his day. He was an unwearied dissector and observer; and it is said, that, on a visitor once asking to see his library, Descartes led him into a room set aside for dissections, and full of specimens under examination. "There," said he, "is my library."

I anticipate a smile of incredulity when I thus champion Descartes' claim to be considered a physiologist of the first rank. I expect to be told that I have read into his works what I find there, and to be asked, Why is it that we are left to discover Descartes' deserts at this time of day, more than two centuries after his death? How is it that Descartes is utterly ignored in some of the latest works which treat expressly of the subject in which he is said to have been so great?

It is much easier to ask such questions than to answer them, especially if one desires to be on good terms with one's contemporaries; but, if I must give an answer, it is this: The growth of physical science is now so prodigiously rapid, that those who are actively engaged in keeping up with the present, have much ado to find time to look at the past, and even grow into the habit of neglecting it. But, natural as this result may be, it is none the less detrimental. The intellect loses, for there is assuredly no more effectual method of clearing up one's own mind on any subject than by talking it over, so to speak, with men of real power and grasp, who have considered it from a totally different point of view. The parallax of time helps us to the true position of a conception, as the parallax of space helps us to that of a star. And the moral nature loses no less. It is well to turn aside from the fretful stir of the present and to dwell with gratitude and respect upon the services of those "mighty men of old who have gone down to the grave with their weapons of war," but who, while they yet lived, won splendid victories over ignorance. It is well, again, to reflect that the fame of Descartes filled all Europe, and his authority overshadowed it, for a century; while now, most of those who know his name think of him, either as a person who had some preposterous notions about vortices and was deservedly annihilated by the great Sir Isaac Newton; or as the apostle of an essentially vicious method of deductive speculation; and that, nevertheless, neither the chatter of shifting opinion, nor the silence of personal oblivion, has in the slightest degree affected the growth of the great ideas of which he was the instrument and the mouthpiece.

It is a matter of fact that the greatest physiologist of the eighteenth century, Haller, in treating of the functions of nerve, does little more than reproduce and enlarge upon the ideas of Descartes. It is a matter of fact that David Hartley, in his remarkable work the "Essay on Man," expressly, though still insufficiently, acknowledges the resemblance of his fundamental conceptions to those of Descartes; and I shall now endeavour to show that a series of propositions, which constitute the foundation and essence of the modern physiology of the nervous system, are fully expressed and illustrated in the works of Descartes.

I. *The brain is the organ of sensation, thought, and emotion; that is to say, some change in the condition of the matter of this organ is the invariable antecedent of the state of consciousness to which each of these terms is applied.*

In the "Principes de la Philosophie" (§ 169), Descartes says:—[1]

"Although the soul is united to the whole body, its principal functions are, nevertheless, performed in the brain; it is here that it not only understands and imagines, but also feels; and this is effected by the intermediation of the nerves, which extend in the form of delicate threads from the brain to all parts of the body, to which they are attached in such a manner, that we can hardly touch any part of the body without setting the extremity of some nerve in motion. This motion passes along the nerve to that part of the brain which is the common sensorium, as I have sufficiently explained in my Treatise on Dioptrics; and the movements which thus travel along the nerves, as far as that part of the brain with which the soul is closely joined and united, cause it, by reason of their diverse characters, to have different thoughts. And it is these different thoughts of the soul, which arise immediately from the movements that are excited by the nerves in the brain, which we properly term our feelings, or the perceptions of our senses."

[1] I quote, here and always, Cousin's edition of the works of Descartes, as most convenient for reference. It is entitled "Œuvres complètes de Descartes," publiées par Victor Cousin. 1824.

Elsewhere,[1] Descartes, in arguing that the seat of the passions is not (as many suppose) the heart, but the brain, uses the following remarkable language :—

"The opinion of those who think that the soul receives its passions in the heart, is of no weight, for it is based upon the fact that the passions cause a change to be felt in that organ; and it is easy to see that this change is felt, as if it were in the heart, only by the intermediation of a little nerve which descends from the brain to it; Just as pain is felt, as if it were in the foot, by the intermediation of the nerves of the foot; and the stars are perceived, as if they were in the heavens, by the intermediation of their light and of the optic nerves. So that it is no more necessary for the soul to exert its functions immediately in the heart, to feel its passions there, than it is necessary that it should be in the heavens to see the stars there."

This definite allocation of all the phenomena of consciousness to the brain as their organ, was a step the value of which it is difficult for us to appraise, so completely has Descartes' view incorporated itself with every-day thought and common language. A lunatic is said to be "crack-brained" or "touched in the head," a confused thinker is "muddle-headed," while a clever man is said to have "plenty of brains;" but it must be remembered that at the end of the last century a considerable, though much over-estimated, anatomist, Bichat, so far from having reached the level of Descartes, could gravely argue that the apparatuses of organic life are the sole seat of the passions, which in no way affect the brain, except so far as it is the agent by which the influence of the passions is transmitted to the muscles.[2]

[1] "Les Passions de l'Âme," Article xxxiii.

[2] "Recherches physiologiques sur la Vie et la Mort." Par Xav. Bichat. Art. Sixième.

Modern physiology, aided by pathology, easily demonstrates that the brain is the seat of all forms of consciousness, and fully bears out Descartes' explanation of the reference of those sensations in the viscera which accompany intense emotion, to these organs. It proves, directly, that those states of consciousness which we call sensations are the immediate consequent of a change in the brain excited by the sensory nerves; and, on the well-known effects of injuries, of stimulants, and of narcotics, it bases the conclusion that thought and emotion are, in like manner, the consequents of physical antecedents.

II. *The movements of animals are due to the change of form of muscles, which shorten and become thicker; and this change of form in a muscle arises from a motion of the substance contained within the nerves which go to the muscle.*

In the "Passions de l'Âme," Art. vii., Descartes writes:—

"Moreover, we know that all the movements of the limbs depend on the muscles, and that these muscles are opposed to one another in such a manner, that when one of them shortens, it draws along the part of the body to which it is attached, and so gives rise to a simultaneous elongation of the muscle which is opposed to it. Then, if it happens, afterwards, that the latter shortens, it causes the former to elongate, and draws towards itself the part to which it is attached. Lastly, we know that all these movements of the muscles, as all the senses, depend on the nerves, which are like little threads or tubes, which all come from the brain, and, like it, contain a certain very subtle air or wind, termed the animal spirits."

The property of muscle mentioned by Descartes now goes by the general name of contractility, but his definition of it remains untouched. The long-continued controversy whether contractile substance, speaking generally, has an inherent power of contraction, or whether it contracts only in virtue of an influence exerted by nerve, is now settled in Haller's favour; but Descartes' statement of the dependence of muscular contraction on nerve holds good for the higher forms of muscle, under normal circumstances; so that, although the structure of the various modifications of contractile matter has been worked out with astonishing minuteness—although the delicate physical and chemical changes which accompany muscular contraction have been determined to an extent of which Descartes could not have dreamed, and have quite upset his hypothesis that the cause of the shortening and thickening of the muscle is the flow of animal spirits into it from the nerves—the important and fundamental part of his statement remains perfectly true.

The like may be affirmed of what he says about nerve. We know now that nerves are not exactly tubes, and that "animal spirits" are myths; but the exquisitely refined methods of investigation of Dubois-Reymond and of Helmholz have no less clearly proved that the antecedent of ordinary muscular contraction is a motion of the molecules of the nerve going to the muscle; and that this motion is propagated with a measurable, and by no means great, velocity, through the substance of the nerve towards the muscle.

With the progress of research, the term "animal spirits" gave way to "nervous fluid," and "nervous fluid" has now given way to "molecular motion of nerve-substance." Our conceptions of what takes place in nerve have altered in the same way as our conceptions of what takes place in a conducting wire have altered, since electricity was shown to be not a fluid, but a mode of molecular motion. The change is of vast importance, but it does not affect Descartes' fundamental idea, that a change in the substance of a motor nerve propagated towards a muscle is the ordinary cause of muscular contraction.

III. *The sensations of animals are due to a motion of the substance of the nerves which connect the sensory organs with the brain.*

In *La Dioptrique* (Discours Quatrième), Descartes explains, more fully than in the passage cited above, his hypothesis of the mode of action of sensory nerves:—

"It is the little threads of which the inner substance of the nerves is composed which subserve sensation. You must conceive that these little threads, being inclosed in tubes, which are always distended and kept open by the animal spirits which they contain, neither press upon nor interfere with one another, and are extended from the brain to the extremities of all the members which are sensitive—in such a manner, that the slightest touch which excites the part of one of the members to which a thread is attached, gives rise to a motion of the part of the brain whence it arises, just as by pulling one of the ends of a stretched cord, the other end is instantaneously moved. . . . And we must take care not to imagine that, in order to feel, the soul needs to behold certain images sent by the objects of sense to the brain, as our philosophers commonly suppose; or, at least, we must conceive these images to be something quite different from what they suppose them to be. For, as all they suppose is that these images ought to resemble the objects which they represent, it is impossible for them to show how they can be formed by the objects received by the organs of the external senses and transmitted to the brain. And they have had no reason for supposing the existence of these images except this; seeing that the mind is readily excited by a picture to conceive the object which is depicted, they have thought that it must be excited in the same way to conceive those objects which affect our senses by little pictures of them formed in the head; instead of which we ought to recollect that there are many things besides images which may excite the mind, as, for example, signs and words, which have not the least resemblance to the objects which they signify." [1]

Modern physiology amends Descartes' conception of the mode of action of sensory nerves in detail, by showing that their structure is the same as that of motor nerves; and that the changes which take place in them, when the sensory organs with which they are connected are excited, are of just the same nature as those which occur in motor nerves, when the muscles to which they are distributed are made to contract: there is a molecular change which, in the case of the sensory nerve, is propagated towards the brain. But the great fact insisted upon by Descartes, that no likeness of external things is, or can be, transmitted to the mind by the sensory organs; on the contrary, that,

[1] Locke (*Human Understanding*, Book II., chap. viii. 37) uses Descartes' illustration for the same purpose, and warns us that "most of the ideas of sensation are no more the likeness of something existing without us than the names that stand for them are the likeness of our ideas, which yet, upon hearing, they are apt to excite in us," a declaration which paved the way for Berkeley.

between the external cause of a sensation and the sensation, there is interposed a mode of motion of nervous matter, of which the state of consciousness is no likeness, but a mere symbol, is of the profoundest importance. It is the physiological foundation of the doctrine of the relativity of knowledge, and a more or less complete idealism is a necessary consequence of it.

For of two alternatives one must be true. Either consciousness is the function of a something distinct from the brain, which we call the soul, and a sensation is the mode in which this soul is affected by the motion of a part of the brain; or there is no soul, and a sensation is something generated by the mode of motion of a part of the brain. In the former case, the phenomena of the senses are purely spiritual affections; in the latter, they are something manufactured by the mechanism of the body, and as unlike the causes which set that mechanism in motion, as the sound of a repeater is unlike the pushing of the spring which gives rise to it.

The nervous system stands between consciousness and the assumed external world, as an interpreter who can talk with his fingers stands between a hidden speaker and a man who is stone deaf—and Realism is equivalent to a belief on the part of the deaf man, that the speaker must also be talking with his fingers. "Les extrêmes se touchent;" the shibboleth of materialists that "thought is a secretion of the brain," is the Fichtean doctrine that "the phenomenal universe is the creation of the Ego," expressed in other language.

IV. *The motion of the matter of a sensory nerve may be transmitted through the brain to motor nerves, and thereby give rise to contraction of the muscles to which these motor nerves are distributed; and this reflection of motion from a sensory into a motor nerve may take place without volition, or even contrary to it.*

In stating these important truths, Descartes defined that which we now term "reflex action." Indeed he almost uses the term itself, as he talks of the "animal spirits" as "réfléchis,"[1] from the sensory into the motor nerves. And that this use of the word "reflected" was no mere accident, but that the importance and appropriateness of the idea it suggests was fully understood by Descartes' contemporaries, is apparent from a passage in Willis's well-known essay, "De Animâ Brutorum," published in 1672, in which, in giving an account of Descartes' views, he speaks of the animal spirits being diverted into motor channels, "velut undulatione reflexâ."[2]

Nothing can be clearer in statement, or in illustration, than the view of reflex action which Descartes gives in the "Passions de l'Âme," Art. xiii.

[1] "Passions de l'Âme," Art. xxxvi.

[2] "Quamcumque Bruti actionem, velut automati mechanici motum artificialem, in eo consistere quod se primò sensibile aliquod spiritus animales afficiens, eosque introrsum convertens, *sensionem* excitat, à qua mox iidem spiritus, velut undulatione reflexâ denuo retrorsum commoti atque pro concinno ipsius fabricæ organorum, et partium ordine, in certos nervos musculosque determinati, respectivos *membrorum motus* perficiunt."—WILLIS: "De Animâ Brutorum," p. 5, ed. 1763.

After recapitulating the manner in which sensory impressions transmitted by the sensory nerves to the brain give rise to sensation, he proceeds :—

"And in addition to the different feelings excited in the soul by these different motions of the brain, the animal spirits, without the intervention of the soul, may take their course towards certain muscles, rather than towards others, and thus move the limbs, as I shall prove by an example. If some one moves his hand rapidly towards our eyes, as if he were going to strike us, although we know that he is a friend, that he does it only in jest, and that he will be very careful to do us no harm, nevertheless it will be hard to keep from winking. And this shows, that it is not by the agency of the soul that the eyes shut, since this action is contrary to that volition which is the only, or at least the chief, function of the soul; but it is because the mechanism of our body is so disposed, that the motion of the hand towards our eyes excites another movement in our brain, and this sends the animal spirits into those muscles which cause the eyelids to close."

Since Descartes' time, experiment has eminently enlarged our knowledge of the details of reflex action. The discovery of Bell has enabled us to follow the tracks of the sensory and motor impulses, along distinct bundles of nerve fibres; and the spinal cord, apart from the brain, has been proved to be a great centre of reflex action; but the fundamental conception remains as Descartes left it, and it is one of the pillars of nerve physiology at the present day.

V. *The motion of any given portion of the matter of the brain excited by the motion of a sensory nerve, leaves behind a readiness to be moved in the same way, in that part. Anything which resuscitates the motion gives rise to the appropriate feeling. This is the physical mechanism of memory.*

Descartes imagined that the pineal body (a curious appendage to the upper side of the brain, the function of which, if it have any, is wholly unknown) was the instrument through which the soul received impressions from, and communicated them to, the brain. And he thus endeavours to explain what happens when one tries to recollect something:—

"Thus when the soul wills to remember anything, this volition, causing the [pineal] gland to incline itself in different directions, drives the [animal] spirits towards different regions of the brain, until they reach that part in which are the traces, which the object which it desires to remember has left. These traces are produced thus: those pores of the brain through which the [animal] spirits have previously been driven, by reason of the presence of the object, have thereby acquired a tendency to be opened by the animal spirits which return towards them, more easily than other pores, so that the animal spirits, impinging on these pores, enter them more readily than others. By this means they excite a particular movement in the pineal gland, which represents the object to the soul, and causes it to know what it is which it desired to recollect."[1]

That memory is dependent upon some condition of the brain is a fact established by many considerations—among the most important of which are the remarkable phenomena of aphasia. And that the condition of the brain on which memory depends, is largely determined by the repeated occurrence of that condition of its molecules, which gives rise to the idea of the thing remembered, is no less certain.

[1] "Les Passions de l'Âme," xlii.

Every boy who learns his lesson by repeating it exemplifies the fact. Descartes, as we have seen, supposes that the pores of a given part of the brain are stretched by the animal spirits, on the occurrence of a sensation, and that the part of the brain thus stretched, being imperfectly elastic, does not return to exactly its previous condition, but remains more distensible than it was before. Hartley supposes that the vibrations, excited by a sensory, or other, impression, do not die away, but are represented by smaller vibrations or "vibratiuncules," the permanency and intensity of which are in relation with the frequency of repetition of the primary vibrations. Haller has substantially the same idea, but contents himself with the general term "mutationes," to express the cerebral change which is the cause of a state of consciousness. These "mutationes" persist for a long time after the cause which gives rise to them has ceased to operate, and are arranged in the brain according to the order of coexistence and succession of their causes. And he gives these persistent "mutationes" the picturesque name of *vestigia rerum*, "quæ non in mente sed in ipso corpore et in medulla quidem cerebri ineffabili modo incredibiliter minutis notis et copia infinita, inscriptæ sunt."[1] I do not know that any modern theory of the physical conditions of memory differs essentially from these, which are all children—*mutatis mutandis*—of the Cartesian doctrine. Physiology is, at present, incompetent to say anything positively about the matter, or to go farther than the expression

[1] Haller, "Primæ Lineæ," ed. iii. "Sensus Interni," dlviii.

of the high probability, that every molecular change which gives rise to a state of consciousness, leaves a more or less persistent structural modification, through which the same molecular change may be regenerated by other agencies than the cause which first produced it.

Thus far, the propositions respecting the physiology of the nervous system which are stated by Descartes have simply been more clearly defined, more fully illustrated, and, for the most part, demonstrated, by modern physiological research. But there remains a doctrine to which Descartes attached great weight, so that full acceptance of it became a sort of note of a thorough-going Cartesian, but which, nevertheless, is so opposed to ordinary prepossessions that it attained more general notoriety, and gave rise to more discussion, than almost any other Cartesian hypothesis. It is the doctrine, that brute animals are mere machines or automata, devoid not only of reason, but of any kind of consciousness, which is stated briefly in the "Discours de la Méthode," and more fully in the "Réponses aux Quatrièmes Objections," and in the correspondence with Henry More.[1]

The process of reasoning by which Descartes arrived at this startling conclusion is well shown in the following passage of the "Réponses:"—

"But as regards the souls of beasts, although this is not the

[1] "Réponse de M. Descartes à M. Morus." 1649. "Œuvres," tome x. p. 204. "Mais le plus grand de tous les préjugés que nous ayons retenus de notre enfance, est celui de croire que les bêtes pensent," etc.

place for considering them, and though, without a general exposition of physics, I can say no more on this subject than I have already said in the fifth part of my Treatise on Method; yet, I will further state, here, that it appears to me to be a very remarkable circumstance that no movement can take place, either in the bodies of beasts, or even in our own, if these bodies have not in themselves all the organs and instruments by means of which the very same movements would be accomplished in a machine. So that, even in us, the spirit, or the soul, does not directly move the limbs, but only determines the course of that very subtle liquid which is called the animal spirits, which, running continually from the heart by the brain into the muscles, is the cause of all the movements of our limbs, and often may cause many different motions, one as easily as the other.

"And it does not even always exert this determination; for among the movements which take place in us, there are many which do not depend on the mind at all, such as the beating of the heart, the digestion of food, the nutrition, the respiration, of those who sleep; and, even in those who are awake, walking, singing, and other similar actions, when they are performed without the mind thinking about them. And, when one who falls from a height throws his hands forwards to save his head, it is in virtue of no ratiocination that he performs this action; it does not depend upon his mind, but takes place merely because his senses being affected by the present danger, some change arises in his brain which determines the animal spirits to pass thence into the nerves, in such a manner as is required to produce this motion, in the same way as in a machine, and without the mind being able to hinder it. Now since we observe this in ourselves, why should we be so much astonished if the light reflected from the body of a wolf into the eye of a sheep has the same force to excite in it the motion of flight?

"After having observed this, if we wish to learn by reasoning, whether certain movements of beasts are comparable to those which are effected in us by the operation of the mind, or, on the contrary, to those which depend only on the animal spirits and the disposition of the organs, it is necessary to consider the difference between the two, which I have explained in the fifth part of the Discourse on Method (for I do not think that any others are discoverable), and then it will easily be seen, that all the actions of beasts are similar only to those which we perform without the help of our minds. For which reason we shall be forced to conclude, that we know of the existence in them of no other principle of motion than the disposition of their organs and the continual affluence of animal spirits produced by the heat of the heart, which attenuates and subtilises the blood; and, at the same time, we shall acknowledge that we have had no reason for assuming any other principle, except that, not having distinguished these two principles of motion, and seeing that the one, which depends only on the animal spirits and the organs, exists in beasts as well as in us, we have hastily concluded that the other, which depends on mind and on thought, was also possessed by them."

Descartes' line of argument is perfectly clear. He starts from reflex action in man, from the unquestionable fact that, in ourselves, co-ordinate, purposive, actions may take place, without the intervention of consciousness or volition, or even contrary to the latter. As actions of a certain degree of complexity are brought about by mere mechanism, why may not actions of still greater complexity be the result of a more refined mechanism? What proof is there that brutes are other than a superior race of marionettes, which eat without pleasure, cry without pain, desire nothing, know nothing, and only simulate intelligence as a bee simulates a mathematician?[1]

The Port Royalists adopted the hypothesis that brutes are machines, and are said to have carried its

[1] Malebranche states the view taken by orthodox Cartesians in 1689 very forcibly: "Ainsi dans les chiens, les chats, et les autres animaux, il n'y a ny intelligence, ny âme spirituelle comme on l'entend ordinairement. Ils mangent sans plaisir; ils crient sans douleur; ils croissent sans le sçavoir; ils ne désirent rien; ils ne connoissent rien; et s'ils agissent avec adresse et d'une manière qui marque l'intelligence,

practical applications so far, as to treat domestic animals with neglect, if not with actual cruelty. As late as the middle of the eighteenth century, the problem was discussed very fully and ably by Bouillier, in his "Essai philosophique sur l'Âme des Bêtes," while Condillac deals with it in his "Traité des Animaux;" but since then it has received little attention. Nevertheless, modern research has brought to light a great multitude of facts, which not only show that Descartes' view is defensible, but render it far more defensible than it was in his day.

It must be premised, that it is wholly impossible absolutely to prove the presence or absence of consciousness in anything but one's own brain, though, by analogy, we are justified in assuming its existence in other men. Now if, by some accident, a man's spinal cord is divided, his limbs are paralysed, so far as his volition is concerned, below the point of injury; and he is incapable of experiencing all those states of consciousness, which, in his uninjured state, would be excited by irritation of those nerves which come off below the injury. If the spinal cord is divided in the middle of the back, for example, the skin of the feet may be cut, or pinched, or burned, or wetted with vitriol, without any sensation of touch, or of pain, arising in consciousness. So far as the man is concerned, therefore, the part of the central nervous

c'est que Dieu les faisant pour les conserver, il a conformé leurs corps de telle manière, qu'ils évitent organiquement, sans le sçavoir, tout ce qui peut les détruire et qu'ils semblent craindre." ("Feuillet de Conches. Méditations Métaphysiques et Correspondance de N. Malebranche. Neuvième Méditation." 1841.)

system which lies beyond the injury is cut off from consciousness. It must indeed be admitted, that, if any one think fit to maintain that the spinal cord below the injury is conscious, but that it is cut off from any means of making its consciousness known to the other consciousness in the brain, there is no means of driving him from his position by logic. But assuredly there is no way of proving it, and in the matter of consciousness, if in anything, we may hold by the rule, "De non apparentibus et de non existentibus eadem est ratio." However near the brain the spinal cord is injured, consciousness remains intact, except that the irritation of parts below the injury is no longer represented by sensation. On the other hand, pressure upon the anterior division of the brain, or extensive injuries to it, abolish consciousness. Hence, it is a highly probable conclusion, that consciousness in man depends upon the integrity of the anterior division of the brain, while the middle and hinder divisions of the brain, and the rest of the nervous centres, have nothing to do with it. And it is further highly probable, that what is true for man is true for other vertebrated animals.

We may assume, then, that in a living vertebrated animal, any segment of the cerebro-spinal axis (or spinal cord and brain) separated from that anterior division of the brain which is the organ of consciousness, is as completely incapable of giving rise to consciousness, as we know it to be incapable of carrying out volitions. Nevertheless, this separated segment of the spinal cord is not passive and inert. On the contrary, it is the seat of extremely remarkable powers. In our imaginary case of injury, the man would, as we have seen, be devoid of sensation in his legs, and would have not the least power of moving them. But, if the soles of his feet were tickled, the legs would be drawn up, just as vigorously as they would have been before the injury. We know exactly what happens when the soles of the feet are tickled; a molecular change takes place in the sensory nerves of the skin, and is propagated along them and through the posterior roots of the spinal nerves, which are constituted by them, to the grey matter of the spinal cord. Through that grey matter, the molecular motion is reflected into the anterior roots of the same nerves, constituted by the filaments which supply the muscles of the legs, and, travelling along these motor filaments, reaches the muscles, which at once contract, and cause the limbs to be drawn up.

In order to move the legs in this way, a definite co-ordination of muscular contractions is necessary; the muscles must contract in a certain order and with duly proportioned force; and moreover, as the feet are drawn away from the source of irritation, it may be said that the action has a final cause, or is purposive.

Thus it follows, that the grey matter of the segment of the man's spinal cord, though it is devoid of consciousness, nevertheless responds to a simple stimulus by giving rise to a complex set of muscular contractions, co-ordinated towards a definite end, and serving an obvious purpose.

If the spinal cord of a frog is cut across, so as to provide us with a segment separated from the brain, we shall have a subject parallel to the injured man, on which experiments can be made without remorse; as we have a right to conclude that a frog's spinal cord is not likely to be conscious, when a man's is not.

Now the frog behaves just as the man did. The legs are utterly paralysed, so far as voluntary movement is concerned; but they are vigorously drawn up to the body when any irritant is applied to the foot. But let us study our frog a little farther. Touch the skin of the side of the body with a little acetic acid, which gives rise to all the signs of great pain in an uninjured frog. In this case, there can be no pain, because the application is made to a part of the skin supplied with nerves which come off from the cord below the point of section; nevertheless, the frog lifts up the limb of the same side, and applies the foot to rub off the acetic acid; and, what is still more remarkable, if the limb be held so that the frog cannot use it, it will, by-and-by, move the limb of the other side, turn it across the body, and use it for the same rubbing process. It is impossible that the frog, if it were in its entirety and could reason, should perform actions more purposive than these: and yet we have most complete assurance that, in this case, the frog is not acting from purpose, has no consciousness, and is a mere insensible machine.

But now suppose that, instead of making a section of the cord in the middle of the body, it had been made in such a manner as to separate the hindermost division of the brain from the rest of the organ, and suppose the foremost two-thirds of the brain entirely taken away. The frog is then absolutely devoid of any spontaneity; it sits upright in the attitude which a frog habitually assumes; and it will not stir unless it is touched; but it differs from the frog which I have just described in this, that, if it be thrown into the water, it begins to swim, and swims just as well as the perfect frog does. But swimming requires the combination and successive co-ordination of a great number of muscular actions. And we are forced to conclude, that the impression made upon the sensory nerves of the skin of the frog by the contact with the water into which it is thrown, causes the transmission to the central nervous apparatus of an impulse, which sets going a certain machinery by which all the muscles of swimming are brought into play in due co-ordination. If the frog be stimulated by some irritating body, it jumps or walks as well as the complete frog can do. The simple sensory impression, acting through the machinery of the cord, gives rise to these complex combined movements.

It is possible to go a step farther. Suppose that only the anterior division of the brain—so much of it as lies in front of the "optic lobes"—is removed. If that operation is performed quickly and skilfully, the frog may be kept in a state of full bodily vigour for months, or it may be for years; but it will sit unmoved. It sees nothing; it hears nothing. It will starve sooner than feed itself, although food put into its mouth is swallowed. On irritation, it jumps or walks; if thrown into the water it swims. If it be put on the hand, it sits there, crouched, perfectly quiet, and would sit there for ever. If the hand be inclined very gently and slowly, so that the frog would naturally tend to slip off, the creature's fore paws are shifted on to the edge of the hand, until he can just prevent himself from falling. If the turning of the hand be slowly continued, he mounts up with great care and deliberation, putting first one leg forward and then another, until he balances himself with perfect precision upon the edge; and, if the turning of the hand is continued, over he goes through the needful set of muscular operations, until he comes to be seated in security, upon the back of the hand. The doing of all this requires a delicacy of co-ordination, and a precision of adjustment of the muscular apparatus of the body, which are only comparable to those of a rope-dancer. To the ordinary influences of light, the frog, deprived of its cerebral hemispheres, appears to be blind. Nevertheless, if the animal be put upon a table, with a book at some little distance between it and the light, and the skin of the hinder part of its body is then irritated, it will jump forward, avoiding the book by passing to the right or left of it. Therefore, although the frog appears to have no sensation of light, visible objects act through its brain upon the motor mechanism of its body.[1]

[1] See the remarkable essay of Göltz, "Beiträge zur Lehre von den Functionen der Nervencentren des Frosches," published in 1869. I have repeated Göltz's experiments, and obtained the same results.

It is obvious, that had Descartes been acquainted with these remarkable results of modern research, they would have furnished him with far more powerful arguments than he possessed in favour of his view of the automatism of brutes. The habits of a frog, leading its natural life, involve such simple adaptations to surrounding conditions, that the machinery which is competent to do so much without the intervention of consciousness, might well do all. And this argument is vastly strengthened by what has been learned in recent times of the marvellously complex operations which are performed mechanically, and to all appearance without consciousness, by men, when, in consequence of injury or disease, they are reduced to a condition more or less comparable to that of a frog, in which the anterior part of the brain has been removed. A case has recently been published by an eminent French physician, Dr. Mesnet, which illustrates this condition so remarkably, that I make no apology for dwelling upon it at considerable length.[1]

A sergeant of the French army, F——, twenty-seven years of age, was wounded during the battle of Bazeilles, by a ball which fractured his left parietal bone. He ran his bayonet through the Prussian soldier who wounded him, but almost immediately his right arm became paralysed; after walking about

[1] "De l'Automatisme de la Mémoire et du Souvenir, dans le Somnambulisme pathologique." Par le Dr. E. Mesnet, Médecin de l'Hôpital Saint-Antoine. "L'Union Médicale," Juillet 21 et 23, 1874. My attention was first called to a summary of this remarkable case, which appeared in the "Journal des Débats" for the 7th of August 1874, by my friend General Strachey, F.R.S.

two hundred yards, his right leg became similarly affected, and he lost his senses. When he recovered them, three weeks afterwards, in hospital at Mayence, the right half of the body was completely paralysed, and remained in this condition for a year. At present, the only trace of the paralysis which remains is a slight weakness of the right half of the body. Three or four months after the wound was inflicted, periodical disturbances of the functions of the brain made their appearance, and have continued ever since. The disturbances last from fifteen to thirty hours; the intervals at which they occur being from fifteen to thirty days.

For four years, therefore, the life of this man has been divided into alternating phases—short abnormal states intervening between long normal states.

In the periods of normal life, the ex-sergeant's health is perfect; he is intelligent and kindly, and performs, satisfactorily, the duties of a hospital attendant. The commencement of the abnormal state is ushered in by uneasiness and a sense of weight about the forehead, which the patient compares to the constriction of a circle of iron; and, after its termination, he complains, for some hours, of dulness and heaviness of the head. But the transition from the normal to the abnormal state takes place in a few minutes, without convulsions or cries, and without anything to indicate the change to a bystander. His movements remain free and his expression calm, except for a contraction of the brow, an incessant movement of the eyeballs, and a chewing motion of the jaws. The eyes are wide open, and their pupils dilated. If the man happens to be in a place to which he is accustomed, he walks about as usual; but, if he is in a new place, or if obstacles are intentionally placed in his way, he stumbles gently against them, stops, and then, feeling over the objects with his hands, passes on one side of them. He offers no resistance to any change of direction which may be impressed upon him, or to the forcible acceleration or retardation of his movements. He eats, drinks, smokes, walks about, dresses and undresses himself, rises and goes to bed at the accustomed hours. Nevertheless, pins may be run into his body, or strong electric shocks sent through it, without causing the least indication of pain; no odorous substance, pleasant or unpleasant, makes the least impression; he eats and drinks with avidity whatever is offered, and takes asafœtida, or vinegar, or quinine, as readily as water; no noise affects him; and light influences him only under certain conditions. Dr. Mesnet remarks, that the sense of touch alone seems to persist, and indeed to be more acute and delicate than in the normal state; and it is by means of the nerves of touch, almost exclusively, that his organism is brought into relation with the external world. Here a difficulty arises. It is clear from the facts detailed, that the nervous apparatus by which, in the normal state, sensations of touch are excited, is that by which external influences determine the movements of the body, in the abnormal state. But does the state of consciousness, which we term a tactile sensation, accompany the operation of this nervous apparatus in the abnormal state? or is consciousness utterly absent, the man being reduced to an insensible mechanism?

It is impossible to obtain direct evidence in favour of the one conclusion or the other; all that can be said is, that the case of the frog shows that the man may be devoid of any kind of consciousness.

A further difficult problem is this. The man is insensible to sensory impressions made through the ear, the nose, the tongue, and, to a great extent, the eye; nor is he susceptible of pain from causes operating during his abnormal state. Nevertheless, it is possible so to act upon his tactile apparatus, as to give rise to those molecular changes in his sensorium, which are ordinarily the causes of associated trains of ideas. I give a striking example of this process in Dr. Mesnet's words:—

"Il se promenait dans le jardin, sous un massif d'arbres, on lui remet à la main sa canne qu'il avait laissé tomber quelques minutes avant. Il la palpe, promène à plusieurs reprises la main sur la poignée coudée de sa canne—devient attentif—semble prêter l'oreille—et, tout-à-coup, appelle 'Henri!' Puis, 'Les voilà! Ils sont au moins une vingtaine! à nous deux, nous en viendrons à bout!' Et alors portant la main derrière son dos comme pour prendre une cartouche, il fait le mouvement de charger son arme, se couche dans l'herbe à plat ventre, la tête cachée par un arbre, dans la position d'un tirailleur, et suit, l'arme épaulée, tous les mouvements de l'ennemi qu'il croit voir à courte distance."

In a subsequent abnormal period, Dr. Mesnet

caused the patient to repeat this scene by placing him in the same conditions. Now, in this case, the question arises whether the series of actions constituting this singular pantomime was accompanied by the ordinary states of consciousness, the appropriate train of ideas, or not? Did the man dream that he was skirmishing? or was he in the condition of one of Vaucauson's automata—a senseless mechanism worked by molecular changes in his nervous system? The analogy of the frog shows that the latter assumption is perfectly justifiable.

The ex-sergeant has a good voice, and had, at one time, been employed as a singer at a café. In one of his abnormal states he was observed to begin humming a tune. He then went to his room, dressed himself carefully, and took up some parts of a periodical novel, which lay on his bed, as if he were trying to find something. Dr. Mesnet, suspecting that he was seeking his music, made up one of these into a roll and put it into his hand. He appeared satisfied, took up his cane and went down-stairs to the door. Here Dr. Mesnet turned him round, and he walked quite contentedly, in the opposite direction, towards the room of the concièrge. The light of the sun shining through a window now happened to fall upon him, and seemed to suggest the footlights of the stage on which he was accustomed to make his appearance. He stopped, opened his roll of imaginary music, put himself into the attitude of a singer, and sang, with perfect execution, three songs, one after the other. After which he wiped his face with his handkerchief and drank, without a grimace, a tumbler of strong vinegar and water which was put into his hand.

An experiment which may be performed upon the frog deprived of the fore part of its brain, well known as Göltz's "Quak-versuch," affords a parallel to this performance. If the skin of a certain part of the back of such a frog is gently stroked with the finger, it immediately croaks. It never croaks unless it is so stroked, and the croak always follows the stroke, just as the sound of a repeater follows the touching of the spring. In the frog, this "song" is innate—so to speak *à priori*—and depends upon a mechanism in the brain governing the vocal apparatus, which is set at work by the molecular change set up in the sensory nerves of the skin of the back by the contact of a foreign body.

In man there is also a vocal mechanism, and the cry of an infant is in the same sense innate and *à priori*, inasmuch as it depends on an organic relation between its sensory nerves and the nervous mechanism which governs the vocal apparatus. Learning to speak, and learning to sing, are processes by which the vocal mechanism is set to new tunes. A song which has been learned has its molecular equivalent, which potentially represents it in the brain, just as a musical box wound up potentially represents an overture. Touch the stop and the overture begins; send a molecular impulse along the proper afferent nerve and the singer begins his song.

Again, the manner in which the frog, though

apparently insensible to light, is yet, under some circumstances, influenced by visual images, finds a singular parallel in the case of the ex-sergeant.

Sitting at a table, in one of his abnormal states, he took up a pen, felt for paper and ink, and began to write a letter to his general, in which he recommended himself for a medal, on account of his good conduct and courage. It occurred to Dr. Mesnet to ascertain experimentally how far vision was concerned in this act of writing. He therefore interposed a screen between the man's eyes and his hands; under these circumstances he went on writing for a short time, but the words became illegible, and he finally stopped, without manifesting any discontent. On the withdrawal of the screen he began to write again where he had left off. The substitution of water for ink in the inkstand had a similar result. He stopped, looked at his pen, wiped it on his coat, dipped it in the water, and began again, with the same effect.

On one occasion, he began to write upon the topmost of ten superimposed sheets of paper. After he had written a line or two, this sheet was suddenly drawn away. There was a slight expression of surprise, but he continued his letter on the second sheet exactly as if it had been the first. This operation was repeated five times, so that the fifth sheet contained nothing but the writer's signature at the bottom of the page. Nevertheless, when the signature was finished, his eyes turned to the top of the blank sheet, and he went through the form of reading over what he had written, a movement of the lips accompanying each word; moreover, with his pen, he put in such corrections as were needed, in that part of the blank page which corresponded with the position of the words which required correction, in the sheets which had been taken away. If the five sheets had been transparent, therefore, they would, when superposed, have formed a properly written and corrected letter.

Immediately after he had written his letter, F—— got up, walked down to the garden, made himself a cigarette, lighted and smoked it. He was about to prepare another, but sought in vain for his tobacco-pouch, which had been purposely taken away. The pouch was now thrust before his eyes and put under his nose, but he neither saw nor smelt it; but, when it was placed in his hand, he at once seized it, made a fresh cigarette, and ignited a match to light the latter. The match was blown out, and another lighted match placed close before his eyes, but he made no attempt to take it; and, if his cigarette was lighted for him, he made no attempt to smoke. All this time the eyes were vacant, and neither winked, nor exhibited any contraction of the pupils. From these and other experiments, Dr. Mesnet draws the conclusion that his patient sees some things and not others; that the sense of sight is accessible to all things which are brought into relation with him by the sense of touch, and, on the contrary, insensible to things which lie outside this relation. He sees the match he holds, and does not see any other.

Just so the frog "sees" the book which is in the

way of his jump, at the same time that isolated visual impressions take no effect upon him.[1]

As I have pointed out, it is impossible to prove that F—— is absolutely unconscious in his abnormal state, but it is no less impossible to prove the contrary; and the case of the frog goes a long way to justify the assumption that, in the abnormal state, the man is a mere insensible machine.

If such facts as these had come under the knowledge of Descartes, would they not have formed an

[1] Those who have had occasion to become acquainted with the phenomena of somnambulism and of mesmerism, will be struck with the close parallel which they present to the proceedings of F. in his abnormal state. But the great value of Dr. Mesnet's observations lies in the fact that the abnormal condition is traceable to a definite injury to the brain, and that the circumstances are such as to keep us clear of the cloud of voluntary and involuntary fictions in which the truth is too often smothered in such cases. In the unfortunate subjects of such abnormal conditions of the brain, the disturbance of the sensory and intellectual faculties is not unfrequently accompanied by a perturbation of the moral nature, which may manifest itself in a most astonishing love of lying for its own sake. And, in this respect, also, F.'s case is singularly instructive, for though, in his normal state, he is a perfectly honest man, in his abnormal condition he is an inveterate thief, stealing and hiding away whatever he can lay hands on, with much dexterity, and with an absurd indifference as to whether the property is his own or not. Hoffman's terrible conception of the "Doppelt-gänger" is realised by men in this state—who live two lives, in the one of which they may be guilty of the most criminal acts, while, in the other, they are eminently virtuous and respectable. Neither life knows anything of the other. Dr. Mesnet states that he has watched a man in his abnormal state elaborately prepare to hang himself, and has let him go on until asphyxia set in, when he cut him down. But on passing into the normal state the would-be suicide was wholly ignorant of what had happened. The problem of responsibility is here as complicated as that of the prince-bishop, who swore as a prince and not as a bishop. "But, highness, if the prince is damned, what will become of the bishop?" said the peasant.

apt commentary upon that remarkable passage in the "Traité de l'Homme," which I have quoted elsewhere,[1] but which is worth repetition ?—

"All the functions which I have attributed to this machine (the body), as the digestion of food, the pulsation of the heart and of the arteries; the nutrition and the growth of the limbs; respiration, wakefulness, and sleep; the reception of light, sounds, odours, flavours, heat, and such like qualities, in the organs of the external senses; the impression of the ideas of these in the organ of common sensation and in the imagination; the retention or the impression of these ideas on the memory: the internal movements of the appetites and the passions; and lastly the external movements of all the limbs, which follow so aptly, as well the action of the objects which are presented to the senses, as the impressions which meet in the memory, that they imitate as nearly as possible those of a real man; I desire, I say, that you should consider that these functions in the machine naturally proceed from the mere arrangement of its organs, neither more nor less than do the movements of a clock, or other automaton, from that of its weights and its wheels; so that, so far as these are concerned, it is not necessary to conceive any other vegetative or sensitive soul, nor any other principle of motion or of life, than the blood and the spirits agitated by the fire which burns continually in the heart, and which is no wise essentially different from all the fires which exist in inanimate bodies."

And would Descartes not have been justified in asking why we need deny that animals are machines, when men, in a state of unconsciousness, perform, mechanically, actions as complicated and as seemingly rational as those of any animals?

But though I do not think that Descartes' hypothesis can be positively refuted, I am not disposed to accept it. The doctrine of continuity is too well established for it to be permissible to me to

[1] "Lay Sermons, Essays and Reviews," p. 355.

suppose that any complex natural phenomenon comes into existence suddenly, and without being preceded by simpler modifications; and very strong arguments would be needed to prove that such complex phenomena, as those of consciousness, first make their appearance in man. We know, that, in the individual man, consciousness grows from a dim glimmer to its full light, whether we consider the infant advancing in years, or the adult emerging from slumber and swoon. We know, further, that the lower animals possess, though less developed, that part of the brain which we have every reason to believe to be the organ of consciousness in man; and as, in other cases, function and organ are proportional, so we have a right to conclude it is with the brain; and that the brutes, though they may not possess our intensity of consciousness, and though, from the absence of language, they can have no trains of thoughts, but only trains of feelings, yet have a consciousness which, more or less distinctly, foreshadows our own.

I confess that, in view of the struggle for existence which goes on in the animal world, and of the frightful quantity of pain with which it must be accompanied, I should be glad if the probabilities were in favour of Descartes' hypothesis; but, on the other hand, considering the terrible practical consequences to domestic animals which might ensue from any error on our part, it is as well to err on the right side, if we err at all, and deal with them as weaker brethren, who are bound, like the rest of us, to pay their toll for living, and suffer what is needful for the general good.

As Hartley finely says, "We seem to be in the place of God to them;" and we may justly follow the precedents He sets in nature in our dealings with them.

But though we may see reason to disagree with Descartes' hypothesis that brutes are unconscious machines, it does not follow that he was wrong in regarding them as automata. They may be more or less conscious, sensitive, automata; and the view that they are such conscious machines is that which is implicitly, or explicitly, adopted by most persons. When we speak of the actions of the lower animals being guided by instinct and not by reason, what we really mean is that, though they feel as we do, yet their actions are the results of their physical organisation. We believe, in short, that they are machines, one part of which (the nervous system) not only sets the rest in motion, and co-ordinates its movements in relation with changes in surrounding bodies, but is provided with special apparatus, the function of which is the calling into existence of those states of consciousness which are termed sensations, emotions, and ideas. I believe that this generally accepted view is the best expression of the facts at present known.

It is experimentally demonstrable—any one who cares to run a pin into himself may perform a sufficient demonstration of the fact—that a mode of motion of the nervous system is the immediate antecedent of a state of consciousness. All but the adherents of "Occasionalism," or of the doctrine of "Pre-established Harmony" (if any such now exist), must admit that we have as much reason for regarding the mode of motion of the nervous system as the cause of the state of consciousness, as we have for regarding any event as the cause of another. How the one phenomenon causes the other we know, as much or as little, as in any other case of causation; but we have as much right to believe that the sensation is an effect of the molecular change, as we have to believe that motion is an effect of impact; and there is as much propriety in saying that the brain evolves sensation, as there is in saying that an iron rod, when hammered, evolves heat.

As I have endeavoured to show, we are justified in supposing that something analogous to what happens in ourselves takes place in the brutes, and that the affections of their sensory nerves give rise to molecular changes in the brain, which again give rise to, or evolve, the corresponding states of consciousness. Nor can there be any reasonable doubt that the emotions of brutes, and such ideas as they possess, are similarly dependent upon molecular brain changes. Each sensory impression leaves behind a record in the structure of the brain—an "ideagenous" molecule, so to speak, which is competent, under certain conditions, to reproduce, in a fainter condition, the state of consciousness which corresponds with that sensory impression; and it is these "ideagenous molecules" which are the physical basis of memory.

It may be assumed, then, that molecular changes in the brain are the causes of all the states of consciousness of brutes. Is there any evidence that these states of consciousness may, conversely, cause those molecular changes which give rise to muscular motion? I see no such evidence. The frog walks, hops, swims, and goes through his gymnastic performances quite as well without consciousness, and consequently without volition, as with it; and, if a frog, in his natural state, possesses anything corresponding with what we call volition, there is no reason to think that it is anything but a concomitant of the molecular changes in the brain which form part of the series involved in the production of motion.

The consciousness of brutes would appear to be related to the mechanism of their body simply as a collateral product of its working, and to be as completely without any power of modifying that working as the steam-whistle which accompanies the work of a locomotive engine is without influence upon its machinery. Their volition, if they have any, is an emotion indicative of physical changes, not a cause of such changes.

This conception of the relations of states of consciousness with molecular changes in the brain—of *psychoses* with *neuroses*—does not prevent us from ascribing free will to brutes. For an agent is free when there is nothing to prevent him from doing that which he desires to do. If a greyhound chases a hare, he is a free agent, because his action is in entire accordance with his strong desire to catch the hare; while so long as he is held back by the leash he is not free, being prevented by external force from following his inclination. And the ascription of freedom to the greyhound under the former circumstances is by no means inconsistent with the other aspect of the facts of the case—that he is a machine impelled to the chase, and caused, at the same time, to have the desire to catch the game by the impression which the rays of light proceeding from the hare make upon his eyes, and through them upon his brain.

Much ingenious argument has, at various times, been bestowed upon the question: How is it possible to imagine that volition, which is a state of consciousness, and, as such, has not the slightest community of nature with matter in motion, can act upon the moving matter of which the body is composed, as it is assumed to do in voluntary acts? But if, as is here suggested, the voluntary acts of brutes—or, in other words, the acts which they desire to perform—are as purely mechanical as the rest of their actions, and are simply accompanied by the state of consciousness called volition, the inquiry, so far as they are concerned, becomes superfluous. Their volitions do not enter into the chain of causation of their actions at all.

The hypothesis that brutes are conscious automata is perfectly consistent with any view that may be held respecting the often discussed and curious question whether they have souls or not; and, if they have souls, whether those souls are immortal or not. It is obviously harmonious with the most literal adherence to the text of Scripture concerning "the beast that perisheth;" but it is not inconsistent with the amiable conviction ascribed by Pope to his "untutored savage," that when he passes to the happy hunting-grounds in the sky, "his faithful dog shall bear him company." If the brutes have consciousness and no souls, then it is clear that, in them, consciousness is a direct function of material changes; while, if they possess immaterial subjects of consciousness, or souls, then, as consciousness is brought into existence only as the consequence of molecular motion of the brain, it follows that it is an indirect product of material changes. The soul stands related to the body as the bell of a clock to the works, and consciousness answers to the sound which the bell gives out when it is struck.

Thus far I have strictly confined myself to the problem with which I proposed to deal at starting—the automatism of brutes. The question is, I believe, a perfectly open one, and I feel happy in running no risk of either Papal or Presbyterian condemnation for the views which I have ventured to put forward. And there are so very few interesting questions which one is, at present, allowed to think out scientifically—to go as far as reason leads, and stop where evidence comes to an end—without speedily being deafened by the tattoo of "the drum ecclesiastic"—that I have luxuriated in my rare freedom, and would now willingly bring this disquisition to an end if I could hope that other people would go no farther. Unfortunately, past experience debars me from entertaining any such hope, even if

> ". . . . . that drum's discordant sound
> Parading round and round and round,"

were not, at present, as audible to me, as it was to the

mild poet who ventured to express his hatred of drums in general, in that well-known couplet.

It will be said, that I mean that the conclusions deduced from the study of the brutes are applicable to man, and that the logical consequences of such application are fatalism, materialism, and atheism—whereupon the drums will beat the *pas de charge.*

One does not do battle with drummers; but I venture to offer a few remarks for the calm consideration of thoughtful persons, untrammelled by foregone conclusions, unpledged to shore-up tottering dogmas, and anxious only to know the true bearings of the case.

It is quite true that, to the best of my judgment, the argumentation which applies to brutes holds equally good of men; and, therefore, that all states of consciousness in us, as in them, are immediately caused by molecular changes of the brain-substance. It seems to me that in men, as in brutes, there is no proof that any state of consciousness is the cause of change in the motion of the matter of the organism. If these positions are well based, it follows that our mental conditions are simply the symbols in consciousness of the changes which take place automatically in the organism; and that, to take an extreme illustration, the feeling we call volition is not the cause of a voluntary act, but the symbol of that state of the brain which is the immediate cause of that act. We are conscious automata, endowed with free will in the only intelligible sense of that much-abused term—inasmuch as in many respects we are able to do as we like—but none the less parts of the great series of causes and effects which, in unbroken continuity, composes that which is, and has been, and shall be—the sum of existence.

As to the logical consequences of this conviction of mine, I may be permitted to remark that logical consequences are the scarecrows of fools and the beacons of wise men. The only question which any wise man can ask himself, and which any honest man will ask himself, is whether a doctrine is true or false. Consequences will take care of themselves; at most their importance can only justify us in testing with extra care the reasoning process from which they result.

So that if the view I have taken did really and logically lead to fatalism, materialism, and atheism, I should profess myself a fatalist, materialist, and atheist; and I should look upon those who, while they believed in my honesty of purpose and intellectual competency, should raise a hue and cry against me, as people who by their own admission preferred lying to truth, and whose opinions therefore were unworthy of the smallest attention.

But, as I have endeavoured to explain on other occasions, I really have no claim to rank myself among fatalistic, materialistic, or atheistic philosophers. Not among fatalists, for I take the conception of necessity to have a logical, and not a physical foundation; not among materialists, for I am utterly incapable of conceiving the existence of matter if there is no mind in which to picture that existence; not among atheists, for the problem of the ultimate cause of existence is one which seems to me to be hopelessly out of reach of my poor powers. Of all the senseless babble I have ever had occasion to read, the demonstrations of these philosophers who undertake to tell us all about the nature of God would be the worst, if they were not surpassed by the still greater absurdities of the philosophers who try to prove that there is no God.

And if this personal disclaimer should not be enough, let me further point out that a great many persons whose acuteness and learning will not be contested, and whose Christian piety, and, in some cases, strict orthodoxy, are above suspicion, have held more or less definitely the view that man is a conscious automaton.

It is held, for example, in substance, by the whole school of predestinarian theologians, typified by St. Augustine, Calvin, and Jonathan Edwards—the great work of the latter on the will showing in this, as in other cases, that the growth of physical science has introduced no new difficulties of principle into theological problems, but has merely given visible body, as it were, to those which already existed.

Among philosophers, the pious Geulincx and the whole school of occasionalist Cartesians held this view; the orthodox Leibnitz invented the term "automate spirituel," and applied it to man; the fervent Christian, Hartley, was one of the chief advocates and best expositors of the doctrine; while another zealous apologist of Christianity in a sceptical age, and a contemporary of Hartley, Charles Bonnet, the Genevese naturalist, has embodied the doctrine in language of such precision and simplicity, that I will quote the little-known passage of his "Essai de Psychologie" at length:—

"ANOTHER HYPOTHESIS CONCERNING THE MECHANISM OF IDEAS.[1]

"Philosophers accustomed to judge of things by that which they are in themselves, and not by their relation to received ideas, would not be shocked if they met with the proposition that the soul is a mere spectator of the movements of its body: that the latter performs of itself all that series of actions which constitutes life: that it moves of itself: that it is the body alone which reproduces ideas, compares and arranges them; which forms reasonings, imagines and executes plans of all kinds, etc. This hypothesis, though perhaps of an excessive boldness, nevertheless deserves some consideration.

"It is not to be denied that Supreme Power could create an automaton which should exactly imitate all the external and internal actions of man.

"I understand by external actions, all those movements which pass under our eyes; I term internal actions, all the motions which in the natural state cannot be observed because they take place in the interior of the body—such as the movements of digestion, circulation, sensation, etc. Moreover, I include in this category the movements which give rise to ideas, whatever be their nature.

"In the automaton which we are considering everything would be precisely determined. Everything would occur according to the rules of the most admirable mechanism: one state would succeed another state, one operation would lead to another operation, according to invariable laws; motion would become alternately cause and effect, effect and cause; reaction would answer to action, and reproduction to production.

---

1 "Essai de Psychologie," chap. xxvii.

"Constructed with definite relations to the activity of the beings which compose the world, the automaton would receive impressions from it, and, in faithful correspondence thereto, it would execute a corresponding series of motions.

"Indifferent towards any determination, it would yield equally to all, if the first impressions did not, so to speak, wind up the machine and decide its operations and its course.

"The series of movements which this automaton could execute would distinguish it from all others formed on the same model, but which, not having been placed in similar circumstances, would not have experienced the same impressions, or would not have experienced them in the same order.

"The senses of the automaton, set in motion by the objects presented to it, would communicate their motion to the brain, the chief motor apparatus of the machine. This would put in action the muscles of the hands and feet, in virtue of their secret connection with the senses. These muscles, alternately contracted and dilated, would approximate or remove the automaton from the objects, in the relation which they would bear to the conservation or the destruction of the machine.

"The motions of perception and sensation which the objects would have impressed on the brain, would be preserved in it by the energy of its mechanism. They would become more vivid according to the actual condition of the automaton, considered in itself and relatively to the objects.

"Words being only the motions impressed on the organ of hearing and that of voice, the diversity of these movements, their combination, the order in which they would succeed one another, would represent judgments, reasoning, and all the operations of the mind.

"A close correspondence between the organs of the senses, either by the opening into one another of their nervous ramifications, or by interposed springs (*ressorts*), would establish such a connection in their working, that, on the occasion of the movements impressed on one of these organs, other movements would be excited, or would become more vivid in some of the other senses.

"Give the automaton a soul which contemplates its movements, which believes itself to be the author of them, which has different volitions on the occasion of the different movements, and you will on this hypothesis construct a man.

"But would this man be free? Can the feeling of our liberty, this feeling which is so clear and so distinct and so vivid as to persuade us that we are the authors of our actions, be conciliated with this hypothesis? If it removes the difficulty which attends the conception of the action of the soul on the body, on the other hand it leaves untouched that which meets us in endeavouring to conceive the action of the body on the soul."

But if Leibnitz, Jonathan Edwards, and Hartley—men who rank among the giants of the world of thought—could see no antagonism between the doctrine under discussion and Christian orthodoxy, is it not just possible that smaller folk may be wrong in making such a coil about "logical consequences"? And, seeing how large a share of this clamour is raised by the clergy of one denomination or another, may I say, in conclusion, that it really would be well if ecclesiastical persons would reflect that ordination, whatever deep-seated graces it may confer, has never been observed to be followed by any visible increase in the learning or the logic of its subject. Making a man a Bishop, or entrusting him with the office of ministering to even the largest of Presbyterian congregations, or setting him up to lecture to a Church congress, really does not in the smallest degree augment such title to respect as his opinions may intrinsically possess. And, when such a man presumes on an authority which was conferred upon him for other purposes, to sit in judgment upon matters his incompetence to deal with which is patent, it is permissible to ignore his sacerdotal pretensions, and to tell him, as one would tell a mere common, unconsecrated, layman: that it is not necessary for any man to occupy himself with problems of this kind unless he so choose; life is filled full enough by the performance of its ordinary and obvious duties. But that, if a man elect to become a judge of these grave questions; still more, if he assume the responsibility of attaching praise or blame to his fellow-men for the conclusions at which they arrive touching them, he will commit a sin more grievous than most breaches of the Decalogue, unless he avoid a lazy reliance upon the information that is gathered by prejudice and filtered through passion, unless he go back to the prime sources of knowledge—the facts of nature, and the thoughts of those wise men who for generations past have been her best interpreters.

X.

ON SENSATION AND THE UNITY OF STRUCTURE OF SENSIFEROUS ORGANS.

THE maxim that metaphysical inquiries are barren of result, and that the serious occupation of the mind with them is a mere waste of time and labour, finds much favour in the eyes of the many persons who pride themselves on the possession of sound common sense; and we sometimes hear it enunciated by weighty authorities, as if its natural consequence, the suppression of such studies, had the force of a moral obligation.

In this case, however, as in some others, those who lay down the law seem to forget that a wise legislator will consider, not merely whether his proposed enactment is desirable, but whether obedience to it is possible. For, if the latter question is answered negatively, the former is surely hardly worth debate.

Here, in fact, lies the pith of the reply to those who would make metaphysics contraband of intellect. Whether it is desirable to place a prohibitory duty upon philosophical speculations or not, it is utterly impossible to prevent the importation of them into the mind. And it is not a little curious to observe that those who most loudly profess to abstain from such commodities are, all the while, unconscious consumers, on a great scale, of one or other of their multitudinous disguises or adulterations. With mouths full of the particular kind of heavily buttered toast which they affect, they inveigh against the eating of plain bread. In truth, the attempt to nourish the human intellect upon a diet which contains no metaphysics is about as hopeful as that of certain Eastern sages to nourish their bodies without destroying life. Everybody has heard the story of the pitiless microscopist, who ruined the peace of mind of one of these mild enthusiasts by showing him the animals moving in a drop of the water with which, in the innocency of his heart, he slaked his thirst; and the unsuspecting devotee of plain common sense may look for as unexpected a shock when the magnifier of severe logic reveals the germs, if not the full-grown shapes, of lively metaphysical postulates rampant amidst his most positive and matter-of-fact notions.

By way of escape from the metaphysical Will-o'-the-wisps generated in the marshes of literature and theology, the serious student is sometimes bidden to betake himself to the solid ground of physical science. But the fish of immortal memory, who threw himself out of the frying-pan into the fire, was not more ill advised than the man who seeks sanctuary from philosophical persecution within the walls of the observatory or of the laboratory. It is said that "metaphysics" owe their name to the fact that, in Aristotle's works, questions of pure philosophy are dealt with immediately after those of physics. If so, the accident is happily symbolical of the essential relations of things; for metaphysical speculation follows as closely upon physical theory as black care upon the horseman.

One need but mention such fundamental, and indeed indispensable, conceptions of the natural philosopher as those of atoms and forces: or that of attraction considered as action at a distance; or that of potential energy; or the antinomies of a vacuum and a plenum; to call to mind the metaphysical background of physics and chemistry; while, in the biological sciences, the case is still worse. What is an individual among the lower plants and animals? Are genera and species realities or abstractions? Is there such a thing as Vital Force? or does the name denote a mere relic of metaphysical fetichism? Is the doctrine of final causes legitimate or illegitimate? These are a few of the metaphysical topics which are suggested by the most elementary study of biological facts. But, more than this, it may be truly said that the roots of every system of philosophy lie deep among the facts of physiology. No one can doubt that the organs and the functions of sensation are as much a part of the province of the physiologist, as are the organs and functions of motion, or those of digestion; and yet it is impossible to gain an acquaintance with even the rudiments of the physiology of sensation without being led straight to one of the most fundamental of all metaphysical problems. In fact, the sensory operations have been, from time immemorial, the battle-ground of philosophers.

I have more than once taken occasion to point out that we are indebted to Descartes, who happened to be a physiologist as well as a philosopher, for the first distinct enunciation of the essential elements of the true theory of sensation. In later times, it is not to the works of the philosophers, if Hartley and James Mill are excepted, but to those of the physiologists, that we must turn for an adequate account of the sensory process. Haller's luminous, though summary, account of sensation in his admirable "Primæ Lineæ," the first edition of which was printed in 1747, offers a striking contrast to the prolixity and confusion of thought which prevade Reid's "Inquiry," of seventeen years' later date.[1] Even Sir William Hamilton, learned historian and acute critic as he was, not only failed to apprehend the philosophical bearing of long-established physiological truths; but, when he affirmed that there is no reason to deny that the mind feels at the finger points, and none to assert that the brain is the sole organ of thought,[2] he showed that

[1] In justice to Reid, however, it should be stated that the chapters on sensation in the "Essays on the Intellectual Powers" (1785) exhibit a great improvement. He is, in fact, in advance of his commentator, as the note to Essay II. chap. ii. p. 248 of Hamilton's edition shows.

[2] Haller, amplifying Descartes, writes in the "Primæ Lineæ," CCCLXVI.—"Non est adeo obscurum sensum omnem oriri áb objecti sensibilis impressione in nervum quemcumque corporis humani, et eamdem per eum nervum ad cerebrum pervenientem tunc demum repraesentari animæ, quando cerebrum adtigit. Ut etiam hoc falsum sit animam inproximo per sensoria nervorumque ramos sentire." . . . DLVII.—"Dum ergo sentimus quinque diversissima entia conjunguntur:

he had not apprehended the significance of the revolution commenced, two hundred years before his time, by Descartes, and effectively followed up by Haller, Hartley, and Bonnet, in the middle of the last century.

In truth, the theory of sensation, except in one point, is, at the present moment, very much where Hartley, led by a hint of Sir Isaac Newton's, left it, when, a hundred and twenty years since, the "Observations on Man: his Frame, his Duty, and his Expectations," was laid before the world. The whole matter is put in a nutshell in the following passages of this notable book.

"External objects impressed upon the senses occasion, first on the nerves on which they are impressed, and then on the brain, vibrations of the small and, as we may say, infinitesimal medullary particles.

"These vibrations are motions backwards and forwards of the small particles; of the same kind with the oscillations of pendulums and the tremblings of the particles of sounding bodies. They must be conceived to be exceedingly short and small, so as not to have the least efficacy to disturb or move the whole bodies of the nerves or brain." [1]

"The white medullary substance of the brain is also the

---

corpus quod sentimus: organi sensorii adfectio ab eo corpore: cerebri adfectio a sensorii percussione nata: in anima nata mutatio: animæ denique conscientia et sensationis adperceptio." Nevertheless, Sir William Hamilton gravely informs his hearers:—"We have no more right to deny that the mind feels at the finger points, as consciousness assures us, than to assert that it thinks exclusively in the brain."—"Lecture on Metaphysics and Logic," ii. p. 128. "We have no reason whatever to doubt the report of consciousness, that we actually perceive at the external point of sensation, and that we perceive the material reality."—*Ibid.* p. 129.

[1] "Observations on Man," vol. i. p. 11.

immediate instrument by which ideas are presented to the mind; or, in other words, whatever changes are made in this substance, corresponding changes are made in our ideas; and *vice versa*."[1]

Hartley, like Haller, had no conception of the nature and functions of the grey matter of the brain. But, if for "white medullary substance," in the latter paragraph, we substitute "grey cellular substance," Hartley's propositions embody the most probable conclusions which are to be drawn from the latest investigations of physiologists. In order to judge how completely this is the case, it will be well to study some simple case of sensation, and, following the example of Reid and of James Mill, we may begin with the sense of smell. Suppose that I become aware of a musky scent, to which the name of "muskiness" may be given. I call this an odour, and I class it along with the feelings of light, colours, sounds, tastes, and the like, among those phenomena which are known as sensations. To say that I am aware of this phenomenon, or that I have it, or that it exists, are simply different modes of affirming the same facts. If I am asked how I know that it exists, I can only reply that its existence and my knowledge of it are one and the same thing; in short, that my knowledge is immediate or intuitive, and, as such, is possessed of the highest conceivable degree of certainty.

[1] *Ibid.* p. 8. The speculations of Bonnet are remarkably similar to those of Hartley; and they appear to have originated independently, though the "Essai de Psychologie" (1754) is of five years' later date than the "Observations on Man" (1749).

The pure sensation of muskiness is almost sure to be followed by a mental state which is not a sensation, but a belief, that there is somewhere close at hand a something on which the existence of the sensation depends. It may be a musk-deer, or a musk-rat, or a musk-plant, or a grain of dry musk, or simply a scented handkerchief; but former experience leads us to believe that the sensation is due to the presence of one or other of these objects, and that it will vanish if the object is removed. In other words, there arises a belief in an external cause of the muskiness, which, in common language, is termed an odorous body.

But the manner in which this belief is usually put into words is strangely misleading. If we are dealing with a musk-plant, for example, we do not confine ourselves to a simple statement of that which we believe, and say that the musk-plant is the cause of the sensation called muskiness; but we say that the plant has a musky smell, and we speak of the odour as a quality, or property, inherent in the plant. And the inevitable reaction of words upon thought has in this case become so complete, and has penetrated so deeply, that when an accurate statement of the case—namely, that muskiness, inasmuch as the term denotes nothing but a sensation, is a mental state, and has no existence except as a mental phenomenon—is first brought under the notice of common-sense folks, it is usually regarded by them as what they are pleased to call a mere metaphysical paradox and a patent example of useless subtlety. Yet the slightest reflection must suffice to convince any one possessed of sound reasoning faculties, that it is as absurd to suppose that muskiness is a quality inherent in one plant, as it would be to imagine that pain is a quality inherent in another, because we feel pain when a thorn pricks the finger.

Even the common-sense philosopher, *par excellence*, says of smell: "It appears to be a simple and original affection or feeling of the mind, altogether inexplicable and unaccountable. It is indeed impossible that it can be in any body: it is a sensation, and a sensation can only be in a sentient thing."[1]

That which is true of muskiness is true of every other odour. Lavender-smell, clove-smell, garlic-smell, are, like "muskiness," names of states of consciousness, and have no existence except as such. But, in ordinary language, we speak of all these odours as if they were independent entities residing in lavender, cloves, and garlic; and it is not without a certain struggle that the false metaphysic of so-called common sense, thus ingrained in us, is expelled.

For the present purpose, it is unnecessary to inquire into the origin of our belief in external bodies,

[1] "An Inquiry into the Human Mind on the Principles of Common Sense," chap. ii. § 2. Reid affirms that "it is genius, and not the want of it, that adulterates philosophy, and fills it with error and false theory;" and no doubt his own lucubrations are free from the smallest taint of the impurity to which he objects. But, for want of something more than that sort of "common sense," which is very common and a little dull, the contemner of genius did not notice that the admission here made knocks so big a hole in the bottom of "common sense philosophy," that nothing can save it from foundering in the dreaded abyss of Idealism.

or into that of the notion of causation. Assuming the existence of an external world, there is no difficulty in obtaining experimental proof that, as a general rule, olfactory sensations are caused by odorous bodies; and we may pass on to the next step of the inquiry—namely, how the odorous body produces the effect attributed to it.

The first point to be noted here is another fact revealed by experience; that the appearance of the sensation is governed, not only by the presence of the odorous substance, but by the condition of a certain part of our corporeal structure, the nose. If the nostrils are closed, the presence of the odorous substance does not give rise to the sensation; while, when they are open, the sensation is intensified by the approximation of the odorous substance to them, and by snuffing up the adjacent air in such a manner as to draw it into the nose. On the other hand, looking at an odorous substance, or rubbing it on the skin, or holding it to the ear, does not awaken the sensation. Thus, it can be readily established by experiment that the perviousness of the nasal passages is, in some way, essential to the sensory function; in fact, that the organ of that function is lodged somewhere in the nasal passages. And, since odorous bodies give rise to their effects at considerable distances, the suggestion is obvious that something must pass from them into the sense organ. What is this "something," which plays the part of an intermediary between the odorous body and the sensory organ?

The oldest speculation about the matter dates back to Democritus and the Epicurean School, and it is to be found fully stated in the fourth book of Lucretius. It comes to this: that the surfaces of bodies are constantly throwing off excessively attenuated films of their own substance: and that these films, reaching the mind, excite the appropriate sensations in it.

Aristotle did not admit the existence of any such material films, but conceived that it was the form of the substance, and not its matter, which affected sense, as a seal impresses wax, without losing anything in the process. While many, if not the majority, of the Schoolmen took up an intermediate position and supposed that a something, which was not exactly either material or immaterial, and which they called an "intentional species," effected the needful communication between the bodily cause of sensation and the mind.

But all these notions, whatever may be said for or against them in general, are fundamentally defective, by reason of an oversight which was inevitable, in the state of knowledge at the time in which they were promulgated. What the older philosophers did not know, and could not know, before the anatomist and the physiologist had done their work, is that, between the external object and that mind in which they supposed the sensation to inhere, there lies a physical obstacle. The sense organ is not a mere passage by which the "tenuia simulacra rerum," or the "intentional species" cast off by objects, or the "forms" of sensible things, pass straight to the mind ; on the contrary, it stands as a firm and impervious barrier, through which no material particle of the world without can make its way to the world within.

Let us consider the olfactory sense organ more nearly. Each of the nostrils leads into a passage completely separated from the other by a partition, and these two passages place the nostrils in free communication with the back of the throat, so that they freely transmit the air passing to the lungs when the mouth is shut, as in ordinary breathing. The floor of each passage is flat, but its roof is a high arch, the crown of which is seated between the orbital cavities of the skull, which serve for the lodgment and protection of the eyes; and it therefore lies behind the apparent limits of that feature which, in ordinary language, is called the nose. From the side walls of the upper and back part of these arched chambers, certain delicate plates of bone project, and these, as well as a considerable part of the partition between the two chambers, are covered by a fine, soft, moist membrane. It is to this "Schneiderian," or olfactory, membrane that odorous bodies must obtain direct access, if they are to give rise to their appropriate sensations; and it is upon the relatively large surface, which the olfactory membrane offers, that we must seek for the seat of the organ of the olfactory sense. The only essential part of that organ consists of a multitude of minute rod-like bodies, set perpendicularly to the surface of the membrane, and forming a part of the cellular coat, or epithelium, which covers the olfactory membrane, as the epidermis covers the skin. In the case of the olfactory sense, there can be no doubt that the Democritic hypothesis, at any rate for such odorous substances as musk, has a good foundation. Infinitesimal particles of musk fly off from the surface of the odorous body, and, becoming diffused through the air, are carried into the nasal passages, and thence into the olfactory chambers, where they come into contact with the filamentous extremities of the delicate olfactory epithelium.

But this is not all. The "mind" is not, so to speak, upon the other side of the epithelium. On the contrary, the inner ends of the olfactory cells are connected with nerve fibres, and these nerve fibres, passing into the cavity of the skull, at length end in a part of the brain, the olfactory sensorium. It is certain that the integrity of each, and the physical inter-connection of all these three structures, the epithelium of the sensory organ, the nerve fibres, and the sensorium, are essential conditions of ordinary sensation. That is to say, the air in the olfactory chambers may be charged with particles of musk; but, if either the epithelium, or the nerve fibres, or the sensorium is injured, or if they are physically disconnected from one another, sensation will not arise. Moreover, the epithelium may be said to be receptive, the nerve fibres transmissive, and the sensorium sensifacient. For, in the act of smelling, the particles of the odorous substance produce a molecular change (which Hartley was in all probability right in terming a vibration) in the epithelium, and this change being transmitted to the nerve fibres, passes along them with a measurable velocity, and, finally reaching the sensorium, is immediately followed by the sensation.

Thus, modern investigation supplies a representative of the Epicurean simulacra in the volatile particles of the musk; but it also gives us the stamp of the particles on the olfactory epithelium, without any transmission of matter, as the equivalent of the Aristotelian "form;" while, finally, the modes of motion of the molecules of the olfactory cells, of the nerve, and of the cerebral sensorium, which are Hartley's vibrations, may stand very well for a double of the "intentional species" of the Schoolmen. And this last remark is not intended merely to suggest a fanciful parallel; for, if the cause of the sensation is, as analogy suggests, to be sought in the mode of motion of the object of sense, then it is quite possible that the particular mode of motion of the object is reproduced in the sensorium; exactly as the diaphragm of a telephone reproduces the mode of motion taken up at its receiving end. In other words, the secondary "intentional species" may be, as the Schoolmen thought the primary one was, the last link between matter and mind.

None the less, however, does it remain true that no similarity exists, nor indeed is conceivable, between the cause of the sensation and the sensation. Attend as closely to the sensations of muskiness, or any other odour, as we will, no trace of extension, resistance, or motion is discernible in them. They have no attribute in common with those which we ascribe to matter; they are, in the strictest sense of the words, immaterial entities.

Thus, the most elementary study of sensation justifies Descartes' position, that we know more of mind than we do of body; that the immaterial world is a firmer reality than the material. For the sensation "muskiness" is known immediately. So long as it persists, it is a part of what we call our thinking selves, and its existence lies beyond the possibility of doubt. The knowledge of an objective or material cause of the sensation, on the other hand, is mediate; it is a belief as contradistinguished from an intuition; and it is a belief which, in any given instance of sensation, may, by possibility, be devoid of foundation. For odours, like other sensations, may arise from the occurrence of the appropriate molecular changes in the nerve or in the sensorium, by the operation of a cause distinct from the affection of the sense organ by an odorous body. Such "subjective" sensations are as real existences as any others, and as distinctly suggest an external odorous object as their cause; but the belief thus generated is a delusion. And, if beliefs are properly termed "testimonies of consciousness," then undoubtedly the testimony of consciousness may be, and often is, untrustworthy.

Another very important consideration arises out of the facts as they are now known. That which, in the absence of a knowledge of the physiology of sensation, we call the cause of the smell, and term the odorous object, is only such, mediately, by reason of its emitting particles which give rise to a mode of motion in the sense organ. The sense organ, again, is only a mediate cause by reason of its producing a molecular change in the nerve fibre; while this last change is also only a mediate cause of sensation, depending, as it does, upon the change which it excites in the sensorium.

The sense organ, the nerve, and the sensorium, taken together, constitute the sensiferous apparatus. They make up the thickness of the wall between the mind, as represented by the sensation "muskiness," and the object, as represented by the particle of musk in contact with the olfactory epithelium.

It will be observed that the sensiferous wall and the external world are of the same nature; whatever it is that constitutes them both is expressible in terms of matter and motion. Whatever changes take place in the sensiferous apparatus are continuous with, and similar to, those which take place in the external world.[1] But, with the sensorium, matter and motion

[1] The following diagrammatic scheme may help to elucidate the theory of sensation :—

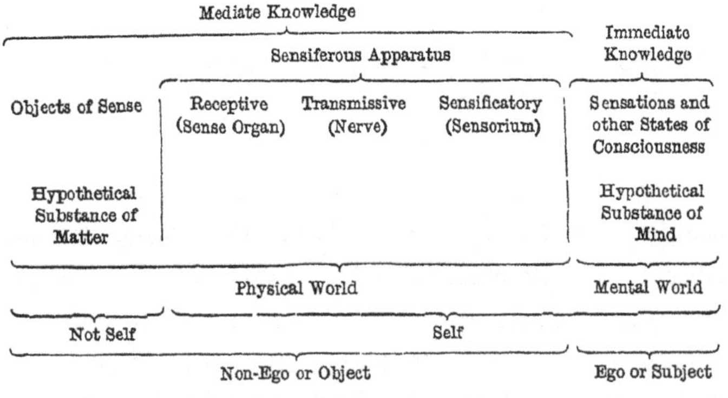

Immediate knowledge is confined to states of consciousness, or, in other

come to an end; while phenomena of another order, or immaterial states of consciousness, make their appearance. How is the relation between the material and the immaterial phenomena to be conceived? This is the metaphysical problem of problems, and the solutions which have been suggested have been made the corner-stones of systems of philosophy. Three mutually irreconcilable readings of the riddle have been offered.

The first is, that an immaterial substance of mind exists; and that it is affected by the mode of motion of the sensorium in such a way as to give rise to the sensation.

The second is, that the sensation is a direct effect of the mode of motion of the sensorium, brought about without the intervention of any substance of mind.

The third is, that the sensation is neither directly nor indirectly an effect of the mode of motion of the sensorium, but that it has an independent cause. Properly speaking, therefore, it is not an effect of the motion of the sensorium, but a concomitant of it.

As none of these hypotheses is capable of even an approximation to demonstration, it is almost needless to remark that they have been severally held with

words, to the phenomena of mind. Knowledge of the physical world, or of one's own body and of objects external to it, is a system of beliefs or judgments based on the sensations. The term "self" is applied not only to the series of mental phenomena which constitute the ego, but to the fragment of the physical world which is their constant concomitant. The corporeal self, therefore, is part of the non-ego; and is objective in relation to the ego as subject.

tenacity and advocated with passion. I do not think it can be said of any of the three that it is inconceivable, or that it can be assumed on *à priori* grounds to be impossible.

Consider the first, for example; an immaterial substance is perfectly conceivable. In fact, it is obvious that, if we possessed no sensations but those of smell and hearing, we should be unable to conceive a material substance. We might have a conception of time, but could have none of extension, or of resistance, or of motion. And without the three latter conceptions no idea of matter could be formed. Our whole knowledge would be limited to that of a shifting succession of immaterial phenomena. But, if an immaterial substance may exist, it may have any conceivable properties; and sensation may be one of them. All these propositions may be affirmed with complete dialectic safety, inasmuch as they cannot possibly be disproved; but neither can a particle of demonstrative evidence be offered in favour of the existence of an immaterial substance.

As regards the second hypothesis, it certainly is not inconceivable, and therefore it may be true, that sensation is the direct effect of certain kinds of bodily motion. It is just as easy to suppose this as to suppose, on the former hypothesis, that bodily motion affects an immaterial substance. But neither is it susceptible of proof.

And, as to the third hypothesis, since the logic of induction is in no case competent to prove that events apparently standing in the relation of cause and effect may not both be effects of a common cause—that also is as safe from refutation, if as incapable of demonstration, as the other two.

In my own opinion, neither of these speculations can be regarded seriously as anything but a more or less convenient working hypothesis. But, if I must choose among them, I take the "law of parcimony" for my guide, and select the simplest—namely, that the sensation is the direct effect of the mode of motion of the sensorium. It may justly be said that this is not the slightest explanation of sensation; but then am I really any the wiser, if I say that a sensation is an activity (of which I know nothing) of a substance of mind (of which also I know nothing)? Or, if I say that the Deity causes the sensation to arise in my mind immediately after He has caused the particles of the sensorium to move in a certain way, is anything gained? In truth, a sensation, as we have already seen, is an intuition—a part of immediate knowledge. As such, it is an ultimate fact and inexplicable; and all that we can hope to find out about it, and that indeed is worth finding out, is its relation to other natural facts. That relation appears to me to be sufficiently expressed, for all practical purposes, by saying that sensation is the invariable consequent of certain changes in the sensorium—or, in other words, that, so far as we know, the change in the sensorium is the cause of the sensation.

I permit myself to imagine that the untutored, if noble, savage of "common sense" who has been misled into reading thus far by the hope of getting positive solid information about sensation, giving way to not unnatural irritation, may here interpellate: "The upshot of all this long disquisition is that we are profoundly ignorant. We knew that to begin with, and you have merely furnished another example of the emptiness and uselessness of metaphysics." But I venture to reply, Pardon me, you were ignorant, but you did not know it. On the contrary, you thought you knew a great deal, and were quite satisfied with the particularly absurd metaphysical notions which you were pleased to call the teachings of common sense. You thought that your sensations were properties of external things, and had an existence outside of yourself. You thought that you knew more about material than you do about immaterial existences. And if, as a wise man has assured us, the knowledge of what we don't know is the next best thing to the knowledge of what we do know, this brief excursion into the province of philosophy has been highly profitable.

Of all the dangerous mental habits, that which schoolboys call "cocksureness" is probably the most perilous; and the inestimable value of metaphysical discipline is that it furnishes an effectual counterpoise to this evil proclivity. Whoso has mastered the elements of philosophy knows that the attribute of unquestionable certainty appertains only to the existence of a state of consciousness so long as it exists; all other beliefs are mere probabilities of a higher or lower order. Sound metaphysic is an amulet which renders its possessor proof alike against the poison of superstition and the counter-poison of nihilism; by showing that the affirmations of the former and the denials of the latter alike deal with matters about which, for lack of evidence, nothing can be either affirmed or denied.

I have dwelt at length upon the nature and origin of our sensations of smell, on account of the comparative freedom of the olfactory sense from the complications which are met with in most of the other senses.

Sensations of taste, however, are generated in almost as simple a fashion as those of smell. In this case, the sense organ is the epithelium which covers the tongue and the palate: and which sometimes, becoming modified, gives rise to peculiar organs termed "gustatory bulbs," in which the epithelial cells elongate and assume a somewhat rod-like form. Nerve fibres connect the sensory organ with the sensorium, and tastes or flavours are states of consciousness caused by the change of molecular state of the latter. In the case of the sense of touch there is often no sense organ distinct from the general epidermis. But many fishes and amphibia exhibit local modifications of the epidermic cells which are sometimes extraordinarily like the gustatory bulbs; more commonly, both in lower and higher animals, the effect of the contact of external bodies is intensified by the development of hair-like filaments, or of true hairs, the bases of which are in immediate relation with the ends of the sensory nerves. Every one must have noticed the extreme delicacy of the sensations produced by the contact of bodies with the ends of the hairs of the head; and the "whiskers" of cats owe their functional importance to the abundant supply of nerves to the follicles in which their bases are lodged. What part, if any, the so-called "tactile corpuscles," "end bulbs," and "Pacinian bodies," play in the mechanism of touch is unknown. If they are sense organs, they are exceptional in character, in so far as they do not appear to be modifications of the epidermis. Nothing is known respecting the organs of those sensations of resistance which are grouped under the head of the muscular sense; nor of the sensations of warmth and cold; nor of that very singular sensation which we call tickling.

In the case of heat and cold, the organism not only becomes affected by external bodies, far more remote than those which affect the sense of smell; but the Democritic hypothesis is obviously no longer permissible. When the direct rays of the sun fall upon the skin, the sensation of heat is certainly not caused by "attenuated films" thrown off from that luminary, but is due to a mode of motion which is transmitted to us. In Aristotelian phrase, it is the form without the matter of the sun which stamps the sense organ; and this, translated into modern language, means nearly the same thing as Hartley's vibrations. Thus we are prepared for what happens in the case of the auditory and the visual senses. For neither the ear, nor the eye, receives anything but the impulses or vibrations originated by sonorous or luminous bodies. Nevertheless, the receptive apparatus still consists of nothing but specially modified epithelial cells. In the labyrinth of the ear of the higher animals, the free ends of these cells terminate in excessively delicate hair-like filaments; while, in the lower forms of auditory organ, its free surface is beset with delicate hairs like those of the surface of the body, and the transmissive nerves are connected with the bases of these hairs. Thus there is an insensible gradation in the forms of the receptive apparatus, from the organ of touch, on the one hand, to those of taste and smell; and, on the other hand, to that of hearing. Even in the case of the most refined of all the sense organs, that of vision, the receptive apparatus departs but little from the general type. The only essential constituent of the visual sense organ is the retina, which forms so small a part of the eyes of the higher animals; and the simplest eyes are nothing but portions of the integument, in which the cells of the epidermis have become converted into glassy rod-like retinal corpuscles. The outer ends of these are turned towards the light; their sides are more or less extensively coated with a dark pigment, and their inner ends are connected with the transmissive nerve fibres. The light, impinging on these visual rods, produces a change in them which is communicated to the nerve fibres, and, being transmitted to the sensorium, gives rise to the sensation—if indeed all animals which possess eyes are endowed with what we understand as sensation.

In the higher animals, a complicated apparatus of

lenses, arranged on the principle of a camera obscura, serves at once to concentrate and to individualise the pencils of light proceeding from external bodies. But the essential part of the organ of vision is still a layer of cells, which have the form of rods with truncated or conical ends. By what seems a strange anomaly, however, the glassy ends of these are turned not towards, but away from, the light; and the latter has to traverse the layer of nervous tissues with which their outer ends are connected, before it can affect them. Moreover, the rods and cones of the vertebrate retina are so deeply seated, and in many respects so peculiar in character, that it appears impossible, at first sight, that they can have anything to do with that epidermis of which gustatory and tactile, and at any rate the lower forms of auditory and visual, organs are obvious modifications.

Whatever be the apparent diversities among the sensiferous apparatuses, however, they share certain common characters. Each consists of a receptive, a transmissive, and a sensificatory portion. The essential part of the first is an epithelium, of the second, nerve fibres, of the third, a part of the brain; the sensation is always the consequence of the mode of motion excited in the receptive, and sent along the transmissive, to the sensificatory part of the sensiferous apparatus. And, in all the senses, there is no likeness whatever between the object of sense, which is matter in motion, and the sensation, which is an immaterial phenomenon.

On the hypothesis which appears to me to be the

most convenient, sensation is a product of the sensiferous apparatus caused by certain modes of motion which are set up in it by impulses from without. The sensiferous apparatuses are, as it were, factories, all of which at the one end receive raw materials of a similar kind—namely, modes of motion—while, at the other, each turns out a special product, the feeling which constitutes the kind of sensation characteristic of it.

Or, to make use of a closer comparison, each sensiferous apparatus is comparable to a musical-box wound up; with as many tunes as there are separate sensations. The object of a simple sensation is the agent which presses down the stop of one of these tunes, and the more feeble the agent, the more delicate must be the mobility of the stop.[1]

But, if this be true, if the recipient part of the sensiferous apparatus is, in all cases, merely a mechanism affected by coarser or finer kinds of material motion, we might expect to find that all sense organs are fundamentally alike, and result from the modification of the same morphological elements. And this is exactly what does result from all recent histological and embryological investigations.

It has been seen that the receptive part of the olfactory apparatus is a slightly modified epithelium, which lines an olfactory chamber deeply seated between the orbits in adult human beings. But, if we trace back the nasal chambers to their origin in the embryo,

[1] "Chaque fibre est une espèce de touche ou de marteau destiné à rendre un certain ton."—Bonnet, "Essai de Psychologie," chap. iv.

we find, that, to begin with, they are mere depressions of the skin of the fore part of the head, lined by a continuation of the general epidermis. These depressions become pits, and the pits, by the growth of the adjacent parts, gradually acquire the position which they finally occupy. The olfactory organ, therefore, is a specially modified part of the general integument.

The human ear would seem to present greater difficulties. For the essential part of the sense organ, in this case, is the membranous labyrinth, a bag of complicated form, which lies buried in the depths of the floor of the skull, and is surrounded by dense and solid bone. Here, however, recourse to the study of development readily unravels the mystery. Shortly after the time when the olfactory organ appears, as a depression of the skin on the side of the fore part of the head, the auditory organ appears as a similar depression on the side of its back part. The depression, rapidly deepening, becomes a small pouch; and then, the communication with the exterior becoming shut off, the pouch is converted into a closed bag, the epithelial lining of which is a part of the general epidermis segregated from the rest. The adjacent tissues, changing first into cartilage and then into bone, enclose the auditory sac in a strong case, in which it undergoes its further metamorphoses; while the drum, the ear bones, and the external ear, are superadded by no less extraordinary modifications of the adjacent parts. Still more marvellous is the history of the development of the organ of vision. In the place of the eye, as in that of the nose and that of the ear, the young embryo presents a depression of the general integument; but, in man and the higher animals, this does not give rise to the proper sensory organ, but only to part of the accessory structures concerned in vision. In fact, this depression, deepening and becoming converted into a shut sac, produces only the cornea, the aqueous humour, and the crystalline lens of the perfect eye.

The retina is added to this by the outgrowth of the wall of a portion of the brain into a sort of bag, or sac, with a narrow neck, the convex bottom of which is turned outwards, or towards the crystalline lens. As the development of the eye proceeds, the convex bottom of the bag becomes pushed in, so that it gradually obliterates the cavity of the sac, the previously convex wall of which becomes deeply concave. The sac of the brain is now like a double nightcap ready for the head, but the place which the head would occupy is taken by the vitreous humour, while the layer of nightcap next it becomes the retina. The cells of this layer which lie farthest from the vitreous humour, or, in other words, bound the original cavity of the sac, are metamorphosed into the rods and cones. Suppose now that the sac of the brain could be brought back to its original form; then the rods and cones would form part of the lining of a side pouch of the brain. But one of the most wonderful revelations of embryology is the proof of the fact that the brain itself is, at its first beginning, merely an infolding of the epidermic layer of the general integument. Hence it follows that the rods and cones of the vertebrate eye are modified epidermic cells, as much as the crystalline cones of the insect or crustacean eye are; and that the inversion of the position of the former in relation to light arises simply from the roundabout way in which the vertebrate retina is developed.

Thus all the higher sense organs start from one foundation, and the receptive epithelium of the eye, or of the ear, is as much modified epidermis as is that of the nose. The structural unity of the sense organs is the morphological parallel to their identity of physiological function, which, as we have seen, is to be impressed by certain modes of motion; and they are fine or coarse, in proportion to the delicacy or the strength of the impulses by which they are to be affected.

In ultimate analysis, then, it appears that a sensation is the equivalent in terms of consciousness for a mode of motion of the matter of the sensorium. But, if inquiry is pushed a stage farther, and the question is asked, What then do we know about matter and motion? there is but one reply possible. All that we know about motion is that it is a name for certain changes in the relations of our visual, tactile, and muscular sensations; and all that we know about matter is that it is the hypothetical substance of physical phenomena—the assumption of the existence of which is as pure a piece of metaphysical speculation as is that of the existence of the substance of mind.

Our sensations, our pleasures, our pains, and the relations of these, make up the sum total of the elements of positive, unquestionable knowledge. We call a large section of these sensations and their relations matter and motion; the rest we term mind and thinking; and experience shows that there is a certain constant order of succession between some of the former and some of the latter.

This is all that just metaphysical criticism leaves of the idols set up by the spurious metaphysics of vulgar common sense. It is consistent either with pure Materialism, or with pure Idealism, but it is neither. For the Idealist, not content with declaring the truth that our knowledge is limited to facts of consciousness, affirms the wholly unprovable proposition that nothing exists beyond these and the substance of mind. And, on the other hand, the Materialist, holding by the truth that, for anything that appears to the contrary, material phenomena are the causes of mental phenomena, asserts his unprovable dogma, that material phenomena and the substance of matter are the sole primary existences.

Strike out the propositions about which neither controversialist does or can know anything, and there is nothing left for them to quarrel about. Make a desert of the Unknowable, and the divine Astræa of philosophic peace will commence her blessed reign.

XI.

EVOLUTION IN BIOLOGY.

IN the former half of the eighteenth century, the term "evolution" was introduced into biological writings, in order to denote the mode in which some of the most eminent physiologists of that time conceived that the generation of living things took place; in opposition to the hypothesis advocated, in the preceding century, by Harvey in that remarkable work[1] which would give him a claim to rank among the founders of biological science, even had he not been the discoverer of the circulation of the blood.

One of Harvey's prime objects is to defend and establish, on the basis of direct observation, the opinion already held by Aristotle; that, in the higher animals at any rate, the formation of the new organism by the process of generation takes place, not suddenly, by simultaneous accretion of rudiments of all, or of the most important, of the organs of the adult; nor by sudden metamorphosis of a formative substance into a miniature of the whole, which subsequently grows; but by *epigenesis*, or successive differentiation of a

[1] The "Exercitationes de Generatione Animalium," which Dr. George Ent extracted from him and published in 1651.

relatively homogeneous rudiment into the parts and structures which are characteristic of the adult.

"Et primò, quidem, quoniam per *epigenesin* sive partium superexorientium additamentum pullum fabricari certum est: quænam pars ante alias omnes exstruatur, et quid de illa ejusque generandi modo observandum veniat, dispiciemus. Ratum sane est et in ovo manifestè apparet quod *Aristoteles* de perfectorum animalium generatione enuntiat: nimirum, non omnes partes simul fieri, sed ordine aliam post aliam; primùmque existere particulam genitalem, cujus virtute postea (tanquam ex principio quodam) reliquæ omnes partes prosiliant. Qualem in plantarum seminibus (fabis, putà, aut glandibus) gemmam sive apicem protuberantem cernimus, totius futuræ arboris principium. *Estque hæc particula velut filius emancipatus seorsumque collocatus, et principium per se vivens; unde postea membrorum ordo describitur; et quæcunque ad absolvendum animal pertinent, disponuntur.*[1] Quoniam enim *nulla pars se ipsam generat; sed postquam generata est, se ipsam jam auget; ideo eam primùm oriri necesse est, quæ principium augendi contineat (sive enim planta, sive animal est, æque omnibus inest quod vim habeat vegetandi, sive nutriendi)*,[2] simulque reliquas omnes partes suo quamque ordine distinguat et formet; proindeque in eadem primogenita particula anima primario inest, sensus, motusque, et totius vitæ auctor et principium." (Exercitatio 51.)

Harvey proceeds to contrast this view with that of the "Medici," or followers of Hippocrates and Galen, who, "badly philosophising," imagined that the brain, the heart, and the liver were simultaneously first generated in the form of vesicles; and, at the same time, while expressing his agreement with Aristotle in the principle of epigenesis, he maintains that it is the blood which is the primal generative part, and not, as Aristotle thought, the heart.

In the latter part of the seventeenth century, the

[1] "De Generatione Animalium," lib ii. cap. x.

[2] "De Generatione," lib. ii. cap. iv.

doctrine of epigenesis, thus advocated by Harvey, was controverted, on the ground of direct observation, by Malpighi, who affirmed that the body of the chick is to be seen in the egg, before the *punctum sanguineum* makes its appearance. But, from this perfectly correct observation a conclusion which is by no means warranted was drawn; namely, that the chick, as a whole, really exists in the egg antecedently to incubation; and that what happens in the course of the latter process is no addition of new parts, "alias post alias natas," as Harvey puts it, but a simple expansion, or unfolding, of the organs which already exist, though they are too small and inconspicuous to be discovered. The weight of Malpighi's observations therefore fell into the scale of that doctrine which Harvey terms *metamorphosis*, in contradistinction to epigenesis.

The views of Malpighi were warmly welcomed, on philosophical grounds, by Leibnitz,[1] who found in them a support to his hypothesis of monads, and by Male-

[1] "Cependant, pour revenir aux formes ordinaires ou aux âmes matérielles, cette durée qu'il leur faut attribuer à la place de celle qu'on avoit attribuée aux atomes pourroit faire douter si elles ne vont pas de corps en corps; ce qui seroit la métempsychose, à peu près comme quelques philosophes ont cru la transmission du mouvement et celle des espèces. Mais cette imagination est bien éloignée de la nature des choses. Il n'y a point de tel passage; et c'est ici où les transformations de Messieurs Swammerdam, Malpighi, et Leewenhoek, qui sont des plus excellens observateurs de notre tems, sont venues à mon secours, et m'ont fait admettre plus aisément, que l'animal, et toute autre substance organisée ne commence point lorsque nous le croyons, et que sa generation apparente n'est qu'une développement et une espèce d'augmentation. Aussi ai je remarqué que l'auteur de la "Recherche de la Verité," M. Regis, M. Hartsoeker, et d'autres habiles hommes n'ont pas été fort éloignés de ce sentiment." Leibnitz, "Système nouveau de la Nature,"

branche;[1] while, in the middle of the eighteenth century, not only speculative considerations, but a great number of new and interesting observations on the phenomena of generation, led the ingenious Bonnet, and Haller,[2] the first physiologist of the age, to adopt, advocate, and extend them.

Bonnet affirms that, before fecundation, the hen's egg contains an excessively minute but complete chick; and that fecundation and incubation simply cause this germ to absorb nutritious matters, which are deposited in the interstices of the elementary structures of which the miniature chick, or germ, is made up. The consequence of this intussusceptive growth is the "development" or "evolution" of the germ into the visible bird. Thus an organised individual (*tout organisé*) "is a composite body consisting of the original, or *elementary*, parts and of the matters which have been associated with them by the aid of nutrition;" so that, if these matters could be extracted from the individual (*tout*), it would, so to speak, become concentrated in a point, and would thus be

1695. The doctrine of "Emboîtement" is contained in the "Considérations sur le principe de vie," 1705; the preface to the "Theodicée," 1710; and the "Principes de la Nature et de la Grace" (§ 6), 1718.

[1] "Il est vrai que la pensée la plus raisonnable et la plus conforme à l'experience sur cette question très difficile de la formation du fœtus; c'est que les enfans sont déja presque tout formés avant même l'action par laquelle ils sont conçus; et que leurs mères ne font que leur donner l'accroissement ordinaire dans le temps de la grossesse." "De la Recherche de la Verité," livre ii. chap. vii. p. 334, 7th ed., 1721.

[2] The writer is indebted to Dr. Allen Thomson for reference to the evidence contained in a note to Haller's edition of Boerhaave's "Prælectiones Academicæ," vol. v. pt. ii. p. 497, published in 1744, that Haller originally advocated epigenesis.

restored to its primitive condition of a *germ;* "just as by extracting from a bone the calcareous substance which is the source of its hardness, it is reduced to its primitive state of gristle or membrane."[1]

"Evolution" and "development" are, for Bonnet, synonymous terms; and since by "evolution" he means simply the expansion of that which was invisible into visibility, he was naturally led to the conclusion, at which Leibnitz had arrived by a different line of reasoning, that no such thing as generation, in the proper sense of the word, exists in nature. The growth of an organic being is simply a process of enlargement, as a particle of dry gelatine may be swelled up by the intussusception of water; its death is a shrinkage, such as the swelled jelly might undergo on desiccation. Nothing really new is produced in the living world, but the germs which develop have existed since the beginning of things; and nothing really dies, but, when what we call death takes place, the living thing shrinks back into its germ state.[2]

1 "Considérations sur les Corps organisés," chap. x.

2 Bonnet had the courage of his opinions, and in the "Palingénésie Philosophique," part vi. chap. iv., he develops a hypothesis which he terms "évolution naturelle;" and which, making allowance for his peculiar views of the nature of generation, bears no small resemblance to what is understood by "evolution" at the present day:—

"Si la volonté divine a créé par un seul Acte l'Universalité des êtres, d'où venoient ces plantes et ces animaux dont Moyse nous decrit la Production au troisieme et au cinquieme jour du renouvellement de notre monde?

"Abuserois-je de la liberté de conjectures si je disois, que les Plantes et les Animaux qui existent aujourd'hui sont parvenus par une sorte d'evolution naturelle des Etres organisés qui peuplaient ce premier Monde, sorti immédiatement des MAINS du CREATEUR? . . .

The two parts of Bonnet's hypothesis, namely, the doctrine that all living things proceed from pre-existing germs, and that these contain, one inclosed within the other, the germs of all future living things, which is the hypothesis of "*emboîtement*;" and the doctrine that every germ contains in miniature all the organs of the adult, which is the hypothesis of evolution or development, in the primary senses of these words, must be carefully distinguished. In fact, while holding firmly by the former, Bonnet more or less modified the latter in his later writings, and, at length, he admits that a "germ" need not be an actual miniature of the organism; but that it may be merely an "original preformation" capable of producing the latter.[1]

But, thus defined, the germ is neither more nor less than the "particula genitalis" of Aristotle, or the "primordium vegetale" or "ovum" of Harvey; and the "evolution" of such a germ would not be distinguishable from "epigenesis."

Supported by the great authority of Haller, the doc-

"Ne supposons que trois révolutions. La Terre vient de sortir des MAINS du CREATEUR. Des causes preparées par sa SAGESSE font développer de toutes parts les Germes. Les Etres organisés commencent à jouir de l'existence. Ils étoient probablement alors bien différens de ce qu'ils sont aujourd'hui. Ils l'etoient autant que ce premier Monde différoit de celui que nous habitons. Nous manquons de moyens pour juger de ces dissemblances, et peut-être que le plus habile Naturaliste qui auroit été placé dans ce premier Monde y auroit entièrement méconnu nos Plantes et nos Animaux."

[1] "Ce mot (germe) ne désignera pas seulement un corps organisé *réduit en petit*; il désignera encore toute espèce de *préformation originelle dont un Tout organique peut résulter comme de son principe immédiat.*"—"Palingénésie Philosophique," part x. chap. ii.

trine of evolution, or development, prevailed throughout the whole of the eighteenth century, and Cuvier appears to have substantially adopted Bonnet's later views, though probably he would not have gone all lengths in the direction of "emboîtement." In a well-known note to Laurillards' "Éloge," prefixed to the last edition of the "Ossemens fossiles," the "radical de l'être" is much the same thing as Aristotle's "particula genitalis" and Harvey's "ovum."[1]

Bonnet's eminent contemporary, Buffon, held nearly the same views with respect to the nature of the germ, and expresses them even more confidently.

"Ceux qui ont cru que le cœur étoit le premier formé, se sont trompés; ceux qui disent que c'est le sang se trompent aussi: tout est formé en même temps. Si l'on ne consulte que l'observation, le poulet se voit dans l'œuf avant qui'il ait été couvé."

"J'ai ouvert une grande quantité d'œufs à differens temps avant et après l'incubation, et je me suis convaincu par mes yeux que le poulet existe en entier dans le milieu de la cicatricule au moment qu'il sort du corps de la poule."[3]

The "moule intérieur" of Buffon is the aggregate of elementary parts which constitute the individual,

[1] "M. Cuvier considérant que tous les êtres organisés sont dérivés de parens, et ne voyant dans la nature aucune force capable de produire l'organisation, croyait à la pré-existence des germes; non pas à la pré-existence d'un être tout formé, puisqu'il est bien évident que ce n'est que par des développemens successifs que l'être acquiert sa forme; mais, si l'on peut s'exprimer ainsi, à la pré-existence du *radical de l'être*, radical qui existe avant que la série des évolutions ne commence, et qui remonte certainement, suivant la belle observation de Bonnet, à plusieurs generations."—Laurillard, "Éloge de Cuvier," note 12.

[2] "Histoire Naturelle," tom. ii. ed. ii. 1750, p. 350.

[3] *Ibid.* p. 351.

and is thus the equivalent of Bonnet's germ,[1] as defined in the passage cited above. But Buffon further imagined that innumerable "molecules organiques" are dispersed throughout the world, and that alimentation consists in the appropriation by the parts of an organism of those molecules which are analogous to them. Growth, therefore, was, on this hypothesis, a process partly of simple evolution, and partly of what has been termed "syngenesis." Buffon's opinion is, in fact, a sort of combination of views, essentially similar to those of Bonnet, with others, somewhat similar to those of the "Medici" whom Harvey condemns. The "molecules organiques" are physical equivalents of Leibnitz's "monads."

It is a striking example of the difficulty of getting people to use their own powers of investigation accurately, that this form of the doctrine of evolution should have held its ground so long; for it was thoroughly and completely exploded, not long after its enunciation, by Caspar Friederich Wolff, who in his "Theoria Generationis," published in 1759, placed the opposite theory of epigenesis upon the secure foundation of fact, from which it has never been displaced. But Wolff had no immediate successors. The school of Cuvier was lamentably deficient in embryologists; and it was only in the course of the first thirty years of the present century, that Prévost and Dumas in France, and, later on, Döllinger, Pander, Von Bär, Rathke, and Remak in Germany, founded modern embryology; while, at the same time,

[1] See particularly Buffon, *l.c.* p. 41.

they proved the utter incompatibility of the hypothesis of evolution, as formulated by Bonnet and Haller, with easily demonstrable facts.

Nevertheless, though the conceptions originally denoted by "evolution" and "development" were shown to be untenable, the words retained their application to the process by which the embryos of living beings gradually make their appearance; and the terms "Development," "Entwickelung," and "Evolutio," are now indiscriminately used for the series of genetic changes exhibited by living beings, by writers who would emphatically deny that "Development" or "Entwickelung" or "Evolutio," in the sense in which these words were usually employed by Bonnet or by Haller, ever occurs.

Evolution, or development, is, in fact, at present employed in biology as a general name for the history of the steps by which any living being has acquired the morphological and the physiological characters which distinguish it. As civil history may be divided into biography, which is the history of individuals, and universal history, which is the history of the human race, so evolution falls naturally into two categories,—the evolution of the individual, and the evolution of the sum of living beings. It will be convenient to deal with the modern doctrine of evolution under these two heads.

I. *The Evolution of the Individual.*

No exception is, at this time, known to the general law, established upon an immense multitude of direct observations, that every living thing is evolved from a particle of matter in which no trace of the distinctive characters of the adult form of that living thing is discernible. This particle is termed a *germ*. Harvey[1] says—

> "Omnibus viventibus primordium insit, ex quo et a quo proveniant. Liceat hoc nobis *primordium vegetale* nominare; nempe substantiam quandam corpoream vitam habentem potentiâ; vel quoddam per se existens, quod aptum sit, in vegetativam formam, ab interno principio operante, mutari. Quale nempe primordium, ovum est et plantarum semen; tale etiam viviparorum conceptus, et insectorum *vermis* ab Aristotele dictus: diversa scilicet diversorum viventium primordia."

The definition of a germ as "matter potentially alive, and having within itself the tendency to assume a definite living form," appears to meet all the requirements of modern science. For, notwithstanding it might be justly questioned whether a germ is not merely potentially, but rather actually, alive, though its vital manifestations are reduced to a minimum, the term "potential" may fairly be used in a sense broad enough to escape the objection. And the qualification of "potential" has the advantage of reminding us that the great characteristic of the germ is not so much what it is, but what it may, under suitable conditions, become. Harvey shared the belief of Aristotle—whose writings he so often quotes, and of whom he speaks as his precursor and model, with the generous respect with which one genuine worker should regard another—that such germs may

[1] "Exercitationes de Generatione." Ex. 62, "Ovum esse primordium commune omnibus animalibus."

arise by a process of "equivocal generation" out of not-living matter; and the aphorism so commonly ascribed to him, "*omne vivum ex ovo*," and which is indeed a fair summary of his reiterated assertions, though incessantly employed against the modern advocates of spontaneous generation, can be honestly so used only by those who have never read a score of pages of the "Exercitationes." Harvey, in fact, believed as implicitly as Aristotle did in the equivocal generation of the lower animals. But, while the course of modern investigation has only brought out into greater prominence the accuracy of Harvey's conception of the nature and mode of development of germs, it has as distinctly tended to disprove the occurrence of equivocal generation, or abiogenesis, in the present course of nature. In the immense majority of both plants and animals, it is certain that the germ is not merely a body in which life is dormant or potential, but that it is itself simply a detached portion of the substance of a pre-existing living body; and the evidence has yet to be adduced which will satisfy any cautious reasoner that "omne vivum ex vivo" is not as well-established a law of the existing course of nature as "omne vivum ex ovo."

In all instances which have yet been investigated, the substance of this germ has a peculiar chemical composition, consisting of at fewest four elementary bodies, viz. carbon, hydrogen, oxygen, and nitrogen, united into the ill-defined compound known as protein, and associated with much water, and very generally, if not always, with sulphur and phosphorus in minute proportions. Moreover, up to the present time, protein is known only as a product and constituent of living matter. Again, a true germ is either devoid of any structure discernible by optical means, or, at most, it is a simple nucleated cell.[1]

In all cases, the process of evolution consists in a succession of changes of the form, structure, and functions of the germ, by which it passes, step by step, from an extreme simplicity, or relative homogeneity, of visible structure, to a greater or less degree of complexity or heterogeneity; and the course of progressive differentiation is usually accompanied by growth, which is effected by intussusception. This intussusception, however, is a very different process from that imagined either by Buffon, or by Bonnet. The substance by the addition of which the germ is enlarged is, in no case, simply absorbed ready-made from the not-living world and packed between the elementary constituents of the germ, as Bonnet imagined; still less does it consist of the "molecules organiques" of Buffon. The new material is, in great measure, not only absorbed but assimilated, so that it becomes part and parcel of the molecular structure of the living body into which it enters. And, so far from the fully developed organism being simply the germ *plus* the nutriment which it has absorbed, it is probable that the adult contains neither in form, nor in substance, more than an inappreciable fraction of the

[1] In some cases of sexless multiplication the germ is a cell-aggregate—if we call germ only that which is already detached from the parent organism.

constituents of the germ, and that it is almost, if not wholly, made up of assimilated and metamorphosed nutriment. In the great majority of cases, at any rate, the full-grown organism becomes what it is by the absorption of not-living matter, and its conversion into living matter of a specific type. As Harvey says (Ex. 45), all parts of the body are nourished "ab eodem succo alibili, aliter aliterque cambiato," "ut plantæ omnes ex eodem communi nutrimento (sive rore seu terræ humore)."

In all animals and plants, above the lowest, the germ is a nucleated cell, using that term in its broadest sense; and the first step in the process of the evolution of the individual is the division of this cell into two or more portions. The process of division is repeated, until the organism, from being unicellular, becomes multicellular. The single cell becomes a cell-aggregate; and it is to the growth and metamorphosis of the cells of the cell-aggregate thus produced, that all the organs and tissues of the adult owe their origin.

In certain animals belonging to every one of the chief groups into which the *Metazoa* are divisible, the cells of the cell-aggregate which results from the process of yelk-division, and which is termed a *morula*, diverge from one another in such a manner as to give rise to a central space, around which they dispose themselves as a coat or envelope; and thus the morula becomes a vesicle filled with fluid, the *planula*. The wall of the planula is next pushed in on one side, or invaginated, whereby it is converted into a double-walled sac with an opening, the *blastopore*, which leads into the cavity lined by the inner wall. This cavity is the primitive alimentary cavity or *archenteron;* the inner, or invaginated, layer is the *hypoblast,* the outer the *epiblast;* and the embryo, in this stage, is termed a *gastrula.* In all the higher animals, a layer of cells makes its appearance between the hypoblast and the epiblast, and is termed the *mesoblast.* In the further course of development, the epiblast becomes the ectoderm or epidermic layer of the body; the hypoblast becomes the epithelium of the middle portion of the alimentary canal; and the mesoblast gives rise to all the other tissues, except the central nervous system, which originates from an ingrowth of the epiblast.

With more or less modification in detail, the embryo has been observed to pass through these successive evolutional stages in sundry Sponges, Cœlenterates, Worms, Echinoderms, Tunicates, Arthropods, Mollusks, and Vertebrates; and there are valid reasons for the belief, that all animals of higher organisation than the *Protozoa* agree in the general character of the early stages of their individual evolution. Each, starting from the condition of a simple nucleated cell, becomes a cell-aggregate; and this passes through a condition which represents the gastrula stage, before taking on the features distinctive of the group to which it belongs. Stated in this form, the "gastræa theory" of Haeckel appears to the present writer to be one of the most important and best founded of recent generalisations. So far as individual plants and animals are concerned, therefore, evolution is not a speculation but a fact; and it takes place by epigenesis.

"Animal . . . per *epigenesin* procreatur, materiam simul attrahit, parat, concoquit, et eâdem utitur; formatur simul et augetur . . . primum futuri corporis concrementum . . . prout augetur, dividitur sensim et distinguitur in partes, non simul omnes, sed alias post alias natas, et ordine quasque suo emergentes."[1]

In these words, by the divination of genius, Harvey, in the seventeenth century, summed up the outcome of the work of all those who, with appliances he could not dream of, are continuing his labours in the nineteenth century.

Nevertheless, though the doctrine of epigenesis, as understood by Harvey, has definitively triumphed over the doctrine of evolution, as understood by his opponents of the eighteenth century, it is not impossible that, when the analysis of the process of development is carried still farther, and the origin of the molecular components of the physically gross, though sensibly minute, bodies which we term germs is traced, the theory of development will approach more nearly to metamorphosis than to epigenesis. Harvey thought that impregnation influenced the female organism as a contagion; and that the blood, which he conceived to be the first rudiment of the germ, arose in the clear fluid of the "colliquamentum" of the ovum by a process of concrescence, as a sort

[1] Harvey, "Exercitationes de Generatione." Ex. 45, "Quænam sit pulli materia et quomodo fiat in Ovo."

of living precipitate. We now know, on the contrary, that the female germ or ovum, in all the higher animals and plants, is a body which possesses the structure of a nucleated cell; that impregnation consists in the fusion of the substance[1] of another more or less modified nucleated cell, the male germ, with the ovum; and that the structural components of the body of the embryo are all derived, by a process of division, from the coalesced male and female germs. Hence it is conceivable, and indeed probable, that every part of the adult contains molecules, derived both from the male and from the female parent; and that, regarded as a mass of molecules, the entire organism may be compared to a web of which the warp is derived from the female and the woof from the male. And each of these may constitute one individuality, in the same sense as the whole organism is one individual, although the matter of the organism has been constantly changing. The primitive male and female molecules may play the part of Buffon's "moules organiques," and mould the assimilated nutriment, each according to its own type, into innumerable new molecules. From this point of view the process, which, in its superficial aspect, is epigenesis, appears in essence, to be evolution, in the modified sense adopted in Bonnet's later writings; and development is merely the expansion of a potential organism or "original preformation" according to fixed laws.

[1] Not yet actually demonstrated in the case of phænogamous plants.

II. *The Evolution of the Sum of Living Beings.*

The notion that all the kinds of animals and plants may have come into existence by the growth and modification of primordial germs is as old as speculative thought; but the modern scientific form of the doctrine can be traced historically to the influence of several converging lines of philosophical speculation and of physical observation, none of which go farther back than the seventeenth century. These are:—

1. The enunciation by Descartes of the conception that the physical universe, whether living or not living, is a mechanism, and that, as such, it is explicable on physical principles.

2. The observation of the gradations of structure, from extreme simplicity to very great complexity, presented by living things, and of the relation of these graduated forms to one another.

3. The observation of the existence of an analogy between the series of gradations presented by the species which compose any great group of animals or plants, and the series of embryonic conditions of the highest members of that group.

4. The observation that large groups of species of widely different habits present the same fundamental plan of structure; and that parts of the same animal or plant, the functions of which are very different, likewise exhibit modifications of a common plan.

5. The observation of the existence of structures, in a rudimentary and apparently useless condition, in one species of a group, which are fully developed and

have definite functions in other species of the same group.

6. The observation of the effects of varying conditions in modifying living organisms.

7. The observation of the facts of geographical distribution.

8. The observation of the facts of the geological succession of the forms of life.

1. Notwithstanding the elaborate disguise which fear of the powers that were led Descartes to throw over his real opinions, it is impossible to read the "Principes de la Philosophie" without acquiring the conviction that this great philosopher held that the physical world and all things in it, whether living or not living, have originated by a process of evolution, due to the continuous operation of purely physical causes, out of a primitive relatively formless matter.[1]

The following passage is especially instructive :—

"Et tant s'en faut que je veuille que l'on croie toutes les choses que j'écrirai, que même je pretends en proposer ici quelques unes que je crois absolument être fausses ; à savoir, je ne doute point que le monde n'ait été créé au commencement avec autant de perfection qu'il en a ; en sorte que le soleil, la terre, la lune, et les étoiles ont été dès lors ; et que la terre n'a pas eu seulement en soi les semences des plantes, mais que les plantes même en ont couvert une partie ; et qu' Adam et Eve n'ont pas été créés enfans mais en âge d'hommes parfaits. La religion chrétienne veut que nous le croyons ainsi, et la raison naturelle nous persuade entièrement cette vérité ; car si nous considérons la toute puissance

[1] As Buffon has well said :—"L'idée de ramener l'explication de tous les phénomènes à des principes mecaniques est assurement grande et belle, ce pas est le plus hardi qu'on peut faire en philosophie, et c'est Descartes qui l'a fait."—*l.c.* p. 50.

de Dieu, nous devons juger que tout ce qu'il a fait a eu dès le commencement toute la perfection qu'il devoit avoir. Mais néanmoins, comme on connôitroit beaucoup mieux quelle a été la nature d'Adam et celle des arbres de Paradis si on avoit examiné comment les enfants se forment peu à peu dans le ventre de leurs mères et comment les plantes sortent de leurs semences, que si on avoit seulement considéré quels ils ont été quand Dieu les a créés : tout de même, nous ferons mieux entendre quelle est généralement la nature de toutes les choses qui sont au monde si nous pouvons imaginer quelques principes qui soient fort intelligibles et fort simples, desquels nous puissions voir clairement que les astres et la terre et enfin tout ce monde visible auroit pu être produit ainsi que de quelques semences (bien que nous sachions qu'il n'a pas été produit en cette façon) que si nous la decrivions seulement comme il est, ou bien comme nous croyons qu'il a été créé. Et parceque je pense avoir trouvé des principes qui sont tels, je tacherai ici de les expliquer."[1]

If we read between the lines of this singular exhibition of force of one kind and weakness of another, it is clear that Descartes believed that he had divined the mode in which the physical universe had been evolved; and the "Traité de l'homme," and the essay "Sur les Passions" afford abundant additional evidence that he sought for, and thought he had found, an explanation of the phenomena of physical life by deduction from purely physical laws.

Spinoza abounds in the same sense, and is as usual perfectly candid—

"Naturæ leges et regulæ, secundum quas omnia fiunt et ex unis formis in alias mutantur, sunt ubique et semper eadem."[2]

Leibnitz's doctrine of continuity necessarily led him in the same direction; and, of the infinite multi-

[1] "Principes de la Philosophie," Troisième partie, § 45.

[2] "Ethices," Pars tertia, Præfatio.

tude of monads with which he peopled the world, each is supposed to be the focus of an endless process of evolution and involution. In the "Protogæa," xxvi., Leibnitz distinctly suggests the mutability of species—

> "Alii mirantur in saxis passim species videri quas vel in orbe cognito, vel saltem in vicinis locis frustra quæras. Ita "Cornua Ammonis," quæ ex nautilorum numero habeantur, passim et forma et magnitudine (nam et pedali diametro aliquando reperiuntur) ab omnibus illis naturis discrepare dicunt, quas præbet mare. Sed quis absconditos ejus recessus aut subterraneas abyssos pervestigavit? quam multa nobis animalia antea ignota offert novus orbis? Et credibile est per magnas illas conversiones etiam animalium species plurimum immutatas."

Thus, in the end of the seventeenth century, the seed was sown which has, at intervals, brought forth recurrent crops of evolutional hypotheses, based, more or less completely, on general reasonings.

Among the earliest of these speculations is that put forward by Benoit de Maillet in his "Telliamed," which, though printed in 1735, was not published until twenty-three years later. Considering that this book was written before the time of Haller, or Bonnet, or Linnæus, or Hutton, it surely deserves more respectful consideration than it usually receives. For De Maillet not only has a definite conception of the plasticity of living things, and of the production of existing species by the modification of their predecessors; but he clearly apprehends the cardinal maxim of modern geological science, that the explanation of the structure of the globe is to be sought in the deductive application to geological phenomena of the principles established inductively by the study of the present course of nature. Somewhat later, Maupertuis[1] suggested a curious hypothesis as to the causes of variation, which he thinks may be sufficient to account for the origin of all animals from a single pair. Robinet[2] followed out much the same line of thought as De Maillet, but less soberly; and Bonnet's speculations in the "Palingénésie," which appeared in 1769, have already been mentioned. Buffon (1753-1778), at first a partisan of the absolute immutability of species, subsequently appears to have believed that larger or smaller groups of species have been produced by the modification of a primitive stock; but he contributed nothing to the general doctrine of evolution.

Erasmus Darwin ("Zoonomia," 1794), though a zealous evolutionist, can hardly be said to have made any real advance on his predecessors; and, notwithstanding that Goethe (1791-4) had the advantage of a wide knowledge of morphological facts, and a true insight into their signification, while he threw all the power of a great poet into the expression of his conceptions, it may be questioned whether he supplied the doctrine of evolution with a firmer scientific basis than it already possessed. Moreover, whatever the value of Goethe's labours in that field, they were not published before 1820, long after evolutionism had

[1] "Système de la Nature." "Essai sur la Formation des Corps Organisés," 1751, xiv.

[2] "Considérations Philosophiques sur la gradation naturelle des formes de l'être; ou les essais de la nature qui apprend à faire l'homme," 1768.

taken a new departure from the works of Treviranus and Lamarck—the first of its advocates who were equipped for their task with the needful large and accurate knowledge of the phenomena of life, as a whole. It is remarkable that each of these writers seems to have been led, independently and contemporaneously, to invent the same name of "Biology" for the science of the phenomena of life; and thus, following Buffon, to have recognised the essential unity of these phenomena, and their contradistinction from those of inanimate nature. And it is hard to say whether Lamarck or Treviranus has the priority in propounding the main thesis of the doctrine of evolution; for though the first volume of Treviranus's "Biologie" appeared only in 1802, he says, in the preface to his later work, the "Erscheinungen und Gesetze des organischen Lebens," dated 1831, that he wrote the first volume of the "Biologie" "nearly five-and-thirty years ago," or about 1796.

Now, in 1794, there is evidence that Lamarck held doctrines which present a striking contrast to those which are to be found in the "Philosophie Zoologique," as the following passages show:—

"685. Quoique mon unique objet dans cet article n'ait été que de traiter de la cause physique de l'entretien de la vie des êtres organiques, malgré cela j'ai osé avancer en débutant, que l'existence de ces êtres étonnants n'appartiennent nullement à la nature; que tout ce qu'on peut entendre par le mot *nature*, ne pouvoit donner la vie, c'est-à-dire, que toutes les qualités de la matière, jointes à toutes les circonstances possibles, et même à l'activité répandue dans l'univers, ne pouvaient point produire un être muni du mouvement organique, capable de reproduire son semblable, et sujet à la mort.

"686. Tous les individus de cette nature, qui existent, proviennent d'individus semblables qui tous ensemble constituent l'espèce entière. Or, je crois qu'il est aussi impossible à l'homme de connôitre la cause physique du premier individu de chaque espèce, que d'assigner aussi physiquement la cause de l'existence de la matière ou de l'univers entier. C'est au moins ce que le résultat de mes connaissances et de mes réflexions me portent à penser. S'il existe beaucoup de variétés produites par l'effet des circonstances, ces variétés ne dénaturent point les espèces ; mais on se trompe, sans doute souvent, en indiquant comme espèce, ce qui n'est que variété ; et alors je sens que cette erreur peut tirer à conséquence dans les raisonnements que l'on fait sur cette matière." [1]

The first three volumes of Treviranus's "Biologie," which contain his general views of evolution, appeared between 1802 and 1805. The "Recherches sur l'organisation des corps vivants," in which the outlines of Lamarck's doctrines are given, was published in 1802; but the full development of his views, in the "Philosophie Zoologique," did not take place until 1809.

The "Biologie" and the "Philosophie Zoologique" are both very remarkable productions, and are still worthy of attentive study, but they fell upon evil times. The vast authority of Cuvier was employed in support of the traditionally respectable hypotheses of special creation and of catastrophism; and the wild speculations of the "Discours sur les Révolutions de la Surface

[1] "Recherches sur les causes des principaux faits physiques," par J. B. Lamarck. Paris. Seconde année de la République. In the preface, Lamarck says that the work was written in 1776, and presented to the Academy in 1780; but it was not published before 1794, and, at that time, it presumably expressed Lamarck's mature views. It would be interesting to know what brought about the change of opinion manifested in the "Recherches sur l'organisation des corps vivants," published only seven years later.

du Globe" were held to be models of sound scientific thinking, while the really much more sober and philosophical hypotheses of the "Hydrogeologie" were scouted. For many years it was the fashion to speak of Lamarck with ridicule, while Treviranus was altogether ignored.

Nevertheless, the work had been done. The conception of evolution was henceforward irrepressible, and it incessantly reappears, in one shape or another,[1] up to the year 1858, when Mr. Darwin and Mr. Wallace published their "Theory of Natural Selection." The "Origin of Species" appeared in 1859; and it is within the knowledge of all whose memories go back to that time, that, henceforward, the doctrine of evolution has assumed a position and acquired an importance which it never before possessed. In the "Origin of Species," and in his other numerous and important contributions to the solution of the problem of biological evolution, Mr. Darwin confines himself to the discussion of the causes which have brought about the present condition of living matter, assuming such matter to have once come into existence. On the other hand, Mr. Spencer[2] and Professor Haeckel[3] have dealt with the whole problem of evolution. The profound and vigorous writings of Mr. Spencer embody the spirit of Descartes in the knowledge of our own day, and may be regarded as the "Principes de la

[1] See the "Historical Sketch" prefixed to the last edition of the "Origin of Species."

[2] "First Principles" and "Principles of Biology," 1860-1864.

[3] "Generelle Morphologie," 1866.

Philosophie" of the nineteenth century; while, whatever hesitation may not unfrequently be felt by less daring minds, in following Haeckel in many of his speculations, his attempt to systematise the doctrine of evolution and to exhibit its influence as the central thought of modern biology, cannot fail to have a far-reaching influence on the progress of science.

If we seek for the reason of the difference between the scientific position of the doctrine of evolution a century ago, and that which it occupies now, we shall find it in the great accumulation of facts, the several classes of which have been enumerated above, under the second to the eighth heads. For those which are grouped under the second to the seventh of these classes, respectively, have a clear significance on the hypothesis of evolution, while they are unintelligible if that hypothesis be denied. And those of the eighth group are not only unintelligible without the assumption of evolution, but can be proved never to be discordant with that hypothesis, while, in some cases, they are exactly such as the hypothesis requires. The demonstration of these assertions would require a volume, but the general nature of the evidence on which they rest may be briefly indicated.

2. The accurate investigation of the lowest forms of animal life, commenced by Leeuwenhoek and Swammerdam, and continued by the remarkable labours of Reaumur, Trembley, Bonnet, and a host of other observers, in the latter part of the seventeenth and the first half of the eighteenth centuries, drew the attention of biologists to the gradation in the complexity of

organisation which is presented by living beings, and culminated in the doctrine of the "échelle des êtres," so powerfully and clearly stated by Bonnet; and, before him, adumbrated by Locke and by Leibnitz. In the then state of knowledge, it appeared that all the species of animals and plants could be arranged in one series; in such a manner that, by insensible gradations, the mineral passed into the plant, the plant into the polype, the polype into the worm, and so, through gradually higher forms of life, to man, at the summit of the animated world.

But, as knowledge advanced, this conception ceased to be tenable in the crude form in which it was first put forward. Taking into account existing animals and plants alone, it became obvious that they fell into groups which were more or less sharply separated from one another; and, moreover, that even the species of a genus can hardly ever be arranged in linear series. Their natural resemblances and differences are only to be expressed by disposing them as if they were branches springing from a common hypothetical centre.

Lamarck, while affirming the verbal proposition that animals form a single series, was forced by his vast acquaintance with the details of zoology to limit the assertion to such a series as may be formed out of the abstractions constituted by the common characters of each group.[1]

[1] "Il s'agit donc de prouver que la série qui constitute l'échelle animale réside essentiellement dans la distribution des masses principales qui la composent et non dans celle des espèces ni même toujours dans celle des genres."—"Phil. Zoologique," chap. v.

Cuvier on anatomical, and Von Baer on embryological grounds, made the further step of proving that, even in this limited sense, animals cannot be arranged in a single series, but that there are several distinct plans of organisation to be observed among them, no one of which, in its highest and most complicated modification, leads to any of the others.

The conclusions enunciated by Cuvier and Von Baer have been confirmed, in principle, by all subsequent research into the structure of animals and plants. But the effect of the adoption of these conclusions has been rather to substitute a new metaphor for that of Bonnet than to abolish the conception expressed by it. Instead of regarding living things as capable of arrangement in one series like the steps of a ladder, the results of modern investigation compel us to dispose them as if they were the twigs and branches of a tree. The ends of the twigs represent individuals, the smallest groups of twigs species, larger groups genera, and so on, until we arrive at the source of all these ramifications of the main branch, which is represented by a common plan of structure. At the present moment, it is impossible to draw up any definition, based on broad anatomical or developmental characters, by which any one of Cuvier's great groups shall be separated from all the rest. On the contrary, the lower members of each tend to converge towards the lower members of all the others. The same may be said of the vegetable world. The apparently clear distinction between flowering and flowerless plants has been broken down by the series of gradations between the two exhibited by the *Lycopodiaceæ*, *Rhizocarpeæ*, and *Gymnospermeæ*. The groups of *Fungi*, *Lichenes*, and *Algæ* have completely run into one another, and, when the lowest forms of each are alone considered, even the animal and vegetable kingdoms cease to have a definite frontier.

If it is permissible to speak of the relations of living forms to one another metaphorically, the similitude chosen must undoubtedly be that of a common root, whence two main trunks, one representing the vegetable and one the animal world, spring; and, each dividing into a few main branches, these subdivide into multitudes of branchlets and these into smaller groups of twigs.

As Lamarck has well said—[1]

"Il n'y a que ceux qui se sont longtemps et fortement occupés de la détermination des espèces, et qui ont consulté de riches collections, qui peuvent savoir jusqu'à quel point les *espèces*, parmi les corps vivants se fondent les unes dans les autres, et qui ont pu se convaincre que, dans les parties où nous voyons des *espèces* isolès, cela n'est ainsi que parcequ'il nous en manque d'autres qui en sont plus voisines et que nous n'avons pas encore recueillies.

"Je ne veux pas dire pour cela que les animaux qui existent forment une série très-simple et partout également nuancée; mais je dis qu'ils forment une série ramense, irrégulièrement graduée et qui n'a point de discontinuité dans ses parties, ou qui, du moins, n'en a toujours pas eu, s'il est vrai que, par suite de quelques espèces perdues, il s'en trouve quelque part. Il en resulte que les *espèces* qui terminent chaque rameau de la série générale tiennent, au moins d'un côté, à d'autres espèces voisines qui se nuancent avec elles. Voilà ce que l'état bien connu des choses me met maintenant à portée de demontrer. Je n'ai

[1] Philosophie Zoologique, première partie, chap. iii.

besoin d'aucune hypothèse ni d'aucune supposition pour cela: j'en atteste tous les naturalistes observateurs."

3. In a remarkable essay[1] Meckel remarks—

"There is no good physiologist who has not been struck by the observation that the original form of all organisms is one and the same, and that out of this one form, all, the lowest as well as the highest, are developed in such a manner that the latter pass through the permanent forms of the former as transitory stages. Aristotle, Haller, Harvey, Kielmeyer, Autenrieth, and many others, have either made this observation incidentally, or, especially the latter, have drawn particular attention to it, and drawn therefrom results of permanent importance for physiology."

Meckel proceeds to exemplify the thesis, that the lower forms of animals represent stages in the course of the development of the higher, with a large series of illustrations.

After comparing the Salamanders and the perennibranchiate *Urodela* with the Tadpoles and the Frogs, and enunciating the law that the more highly any animal is organised the more quickly does it pass through the lower stages, Meckel goes on to say—

"From these lowest Vertebrata to the highest, and to the highest forms among these, the comparison between the embryonic conditions of the higher animals and the adult states of the lower can be more completely and thoroughly instituted than if the survey is extended to the Invertebrata, inasmuch as the latter are in many respects constructed upon an altogether too dissimilar type; indeed they often differ from one another far more than the lowest vertebrate does from the highest mammal; yet the following pages will show that the comparison may also

---

[1] "Entwurf einer Darstellung der zwischen dem Embryozustände der höheren Thiere und dem permanenten der niederen stattfindenden Parallele," "Beyträge zur Vergleichenden Anatomie," Bd. ii. 1811.

be extended to them with interest. In fact, there is a period when, as Aristotle long ago said, the embryo of the highest animal has the form of a mere worm; and, devoid of internal and external organisation, is merely an almost structureless lump of polype-substance. Notwithstanding the origin of organs, it still for a certain time, by reason of its want of an internal bony skeleton, remains worm and mollusk, and only later enters into the series of the Vertebrata, although traces of the vertebral column even in the earliest periods testify its claim to a place in that series."—*Op. cit.* pp. 4, 5.

If Meckel's proposition is so far qualified, that the comparison of adult with embryonic forms is restricted within the limits of one type of organisation; and, if it is further recollected that the resemblance between the permanent lower form and the embryonic stage of a higher form is not special but general, it is in entire accordance with modern embryology; although there is no branch of biology which has grown so largely, and improved its methods so much, since Meckel's time, as this. In its original form, the doctrine of "arrest of development," as advocated by Geoffroy Saint-Hilaire and Serres, was no doubt an over-statement of the case. It is not true, for example, that a fish is a reptile arrested in its development, or that a reptile was ever a fish: but it is true that the reptile embryo, at one stage of its development, is an organism which, if it had an independent existence, must be classified among fishes; and all the organs of the reptile pass, in the course of their development, through conditions which are closely analogous to those which are permanent in some fishes.

4. That branch of biology which is termed Mor-

phology is a commentary upon, and expansion of, the proposition that widely different animals or plants, and widely different parts of animals or plants, are constructed upon the same plan. From the rough comparison of the skeleton of a bird with that of a man by Belon, in the sixteenth century (to go no farther back), down to the theory of the limbs and the theory of the skull at the present day; or, from the first demonstration of the homologies of the parts of a flower by C. F. Wolff, to the present elaborate analysis of the floral organs, morphology exhibits a continual advance towards the demonstration of a fundamental unity among the seeming diversities of living structures. And this demonstration has been completed by the final establishment of the cell theory, which involves the admission of a primitive conformity, not only of all the elementary structures in animals and plants respectively, but of those in the one of these great divisions of living things with those in the other. No *à priori* difficulty can be said to stand in the way of evolution, when it can be shown that all animals and all plants proceed by modes of development, which are similar in principle, from a fundamental protoplasmic material.

5. The innumerable cases of structures, which are rudimentary and apparently useless, in species, the close allies of which possess well developed and functionally important homologous structures, are readily intelligible on the theory of evolution, while it is hard to conceive their *raison d'être* on any other hypothesis. However, a cautious reasoner will probably rather

explain such cases deductively from the doctrine of evolution than endeavour to support the doctrine of evolution by them. For it is almost impossible to prove that any structure, however rudimentary, is useless—that is to say, that it plays no part whatever in the economy; and, if it is in the slightest degree useful, there is no reason why, on the hypothesis of direct creation, it should not have been created. Nevertheless, double-edged as is the argument from rudimentary organs, there is probably none which has produced a greater effect in promoting the general acceptance of the theory of evolution.

6. The older advocates of evolution sought for the causes of the process exclusively in the influence of varying conditions, such as climate and station, or hybridisation, upon living forms. Even Treviranus has got no farther than this point. Lamarck introduced the conception of the action of an animal on itself as a factor in producing modification. Starting from the well-known fact that the habitual use of a limb tends to develop the muscles of the limb, and to produce a greater and greater facility in using it, he made the general assumption that the effort of an animal to exert an organ in a given direction tends to develop the organ in that direction. But a little consideration showed that, though Lamarck had seized what, as far as it goes, is a true cause of modification, it is a cause the actual effects of which are wholly inadequate to account for any considerable modification in animals, and which can have no influence at all in the vegetable world; and probably nothing con-

tributed so much to discredit evolution, in the early part of this century, as the floods of easy ridicule which were poured upon this part of Lamarck's speculation. The theory of natural selection, or survival of the fittest, was suggested by Wells in 1813, and further elaborated by Matthew in 1831. But the pregnant suggestions of these writers remained practically unnoticed and forgotten, until the theory was independently devised and promulgated by Darwin and Wallace in 1858, and the effect of its publication was immediate and profound.

Those who were unwilling to accept evolution, without better grounds than such as are offered by Lamarck, or the author of that particularly unsatisfactory book, the "Vestiges of the Natural History of the Creation," and who therefore preferred to suspend their judgment on the question, found, in the principle of selective breeding, pursued in all its applications with marvellous knowledge and skill by Mr. Darwin, a valid explanation of the occurrence of varieties and races; and they saw clearly that, if the explanation would apply to species, it would not only solve the problem of their evolution, but that it would account for the facts of teleology, as well as for those of morphology; and for the persistence of some forms of life unchanged through long epochs of time, while others undergo comparatively rapid metamorphosis.

How far "natural selection" suffices for the production of species remains to be seen. Few can doubt that, if not the whole cause, it is a very important factor in that operation; and that it must play a great part in the sorting out of varieties into those which are transitory and those which are permanent.

But the causes and conditions of variation have yet to be thoroughly explored; and the importance of natural selection will not be impaired, even if further inquiries should prove that variability is definite, and is determined in certain directions rather than in others, by conditions inherent in that which varies. It is quite conceivable that every species tends to produce varieties of a limited number and kind, and that the effect of natural selection is to favour the development of some of these, while it opposes the development of others along their predetermined lines of modification.

7. No truths brought to light by biological investigation were better calculated to inspire distrust of the dogmas intruded upon science in the name of theology, than those which relate to the distribution of animals and plants on the surface of the earth. Very skilful accommodation was needful, if the limitation of sloths to South America, and of the ornithorhynchus to Australia, was to be reconciled with the literal interpretation of the history of the deluge; and, with the establishment of the existence of distinct provinces of distribution, any serious belief in the peopling of the world by migration from Mount Ararat came to an end.

Under these circumstances, only one alternative was left for those who denied the occurrence of evolution—namely, the supposition that the characteristic animals and plants of each great province were created, as such, within the limits in which we find them. And as the hypothesis of "specific centres," thus formulated, was heterodox from the theological point of view, and unintelligible under its scientific aspect, it may be passed over without further notice, as a phase of transition from the creational to the evolutional hypothesis.

8. In fact, the strongest and most conclusive arguments in favour of evolution are those which are based upon the facts of geographical, taken in conjunction with those of geological, distribution.

Both Mr. Darwin and Mr. Wallace lay great stress on the close relation which obtains between the existing fauna of any region and that of the immediately antecedent geological epoch in the same region; and rightly, for it is in truth inconceivable that there should be no genetic connection between the two. It is possible to put into words the proposition that all the animals and plants of each geological epoch were annihilated, and that a new set of very similar forms was created for the next epoch; but it may be doubted if any one who ever tried to form a distinct mental image of this process of spontaneous generation on the grandest scale, ever really succeeded in realising it.

Within the last twenty years, the attention of the best palæontologists has been withdrawn from the hodman's work of making "new species" of fossils, to the scientific task of completing our knowledge of individual species, and tracing out the succession of the forms presented by any given type in time.

Those who desire to inform themselves of the

nature and extent of the evidence bearing on these questions may consult the works of Rutimeyer, Gaudry, Kowalewsky, Marsh, and the writer of the present article. It must suffice, in this place, to say that the successive forms of the Equine type have been fully worked out; while those of nearly all the other existing types of Ungulate mammals and of the *Carnivora* have been almost as closely followed through the Tertiary deposits; the gradations between birds and reptiles have been traced; and the modifications undergone by the *Crocodilia*, from the Triassic epoch to the present day, have been demonstrated. On the evidence of palæontology, the evolution of many existing forms of animal life from their predecessors is no longer an hypothesis, but an historical fact; it is only the nature of the physiological factors to which that evolution is due which is still open to discussion.

XII.

THE COMING OF AGE OF "THE ORIGIN OF SPECIES."

MANY of you will be familiar with the aspect of this small green-covered book. It is a copy of the first edition of the "Origin of Species," and bears the date of its production—the 1st of October 1859. Only a few months, therefore, are needed to complete the full tale of twenty-one years since its birthday.

Those whose memories carry them back to this time will remember that the infant was remarkably lively, and that a great number of excellent persons mistook its manifestations of a vigorous individuality for mere naughtiness; in fact there was a very pretty turmoil about its cradle. My recollections of the period are particularly vivid; for, having conceived a tender affection for a child of what appeared to me to be such remarkable promise, I acted for some time in the capacity of a sort of under-nurse, and thus came in for my share of the storms which threatened the very life of the young creature. For some years it was undoubtedly warm work; but considering how exceedingly unpleasant the apparition of the new-comer must have been to those who did not fall in love with him at first sight, I think it is to the credit of our age that the war was not fiercer, and that the more bitter and unscrupulous forms of opposition died away as soon as they did.

I speak of this period as of something past and gone, possessing merely an historical, I had almost said an antiquarian interest. For, during the second decade of the existence of the "Origin of Species," opposition, though by no means dead, assumed a different aspect. On the part of all those who had any reason to respect themselves, it assumed a thoroughly respectful character. By this time, the dullest began to perceive that the child was not likely to perish of any congenital weakness or infantile disorder, but was growing into a stalwart personage, upon whom mere goody scoldings and threatenings with the birch-rod were quite thrown away.

In fact, those who have watched the progress of science within the last ten years will bear me out to the full, when I assert that there is no field of biological inquiry in which the influence of the "Origin of Species" is not traceable; the foremost men of science in every country are either avowed champions of its leading doctrines, or at any rate abstain from opposing them; a host of young and ardent investigators seek for and find inspiration and guidance in Mr. Darwin's great work; and the general doctrine of evolution, to one side of which it gives expression, obtains, in the phenomena of biology, a firm base of operations whence it may conduct its conquest of the whole realm of nature.

History warns us, however, that it is the customary fate of new truths to begin as heresies and to end as superstitions; and, as matters now stand, it is hardly rash to anticipate that, in another twenty years, the new generation, educated under the influences of the present day, will be in danger of accepting the main doctrines of the "Origin of Species," with as little reflection, and it may be with as little justification, as so many of our contemporaries, twenty years ago, rejected them.

Against any such a consummation let us all devoutly pray; for the scientific spirit is of more value than its products, and irrationally held truths may be more harmful than reasoned errors. Now the essence of the scientific spirit is criticism. It tells us that whenever a doctrine claims our assent we should reply, Take it if you can compel it. The struggle for existence holds as much in the intellectual as in the physical world. A theory is a species of thinking, and its right to exist is coextensive with its power of resisting extinction by its rivals.

From this point of view, it appears to me that it would be but a poor way of celebrating the Coming of Age of the "Origin of Species," were I merely to dwell upon the facts, undoubted and remarkable as they are, of its far-reaching influence and of the great following of ardent disciples who are occupied in spreading and developing its doctrines. Mere insanities and inanities have before now swollen to portentous size in the course of twenty years. Let us rather ask this prodigious change in opinion to justify itself; let us inquire whether anything has happened since 1859, which will explain, on rational grounds, why so many are worshipping that which they burned, and burning that which they worshipped. It is only in this way that we shall acquire the means of judging whether the movement we have witnessed is a mere eddy of fashion, or truly one with the irreversible current of intellectual progress, and, like it, safe from retrogressive reaction.

Every belief is the product of two factors: the first is the state of the mind to which the evidence in favour of that belief is presented; and the second is the logical cogency of the evidence itself. In both these respects, the history of biological science during the last twenty years appears to me to afford an ample explanation of the change which has taken place; and a brief consideration of the salient events of that history will enable us to understand why, if the "Origin of Species" appeared now, it would meet with a very different reception from that which greeted it in 1859.

One-and-twenty years ago, in spite of the work commenced by Hutton and continued with rare skill and patience by Lyell, the dominant view of the past history of the earth was catastrophic. Great and sudden physical revolutions, wholesale creations and extinctions of living beings, were the ordinary machinery of the geological epic brought into fashion by the misapplied genius of Cuvier. It was gravely maintained and taught that the end of every geological epoch was signalised by a cataclysm, by which every living being on the globe was swept away, to be replaced by a brand-new creation when the world returned to quiescence. A scheme of nature which appeared to be modelled on the likeness of a succession of rubbers of whist, at the end of each of which the players upset the table and called for a new pack, did not seem to shock anybody.

I may be wrong, but I doubt if, at the present time, there is a single responsible representative of these opinions left. The progress of scientific geology has elevated the fundamental principle of uniformitarianism, that the explanation of the past is to be sought in the study of the present, into the position of an axiom; and the wild speculations of the catastrophists, to which we all listened with respect a quarter of a century ago, would hardly find a single patient hearer at the present day. No physical geologist now dreams of seeking, outside the range of known natural causes, for the explanation of anything that happened millions of years ago, any more than he would be guilty of the like absurdity in regard to current events.

The effect of this change of opinion upon biological speculation is obvious. For, if there have been no periodical general physical catastrophes, what brought about the assumed general extinctions and re-creations of life which are the corresponding biological catastrophes? And, if no such interruptions of the ordinary course of nature have taken place in the organic, any more than in the inorganic, world, what alternative is there to the admission of evolution?

The doctrine of evolution in biology is the necessary result of the logical application of the principles of uniformitarianism to the phenomena of life. Darwin is the natural successor of Hutton and Lyell, and the "Origin of Species" the logical sequence of the "Principles of Geology."

The fundamental doctrine of the "Origin of Species," as of all forms of the theory of evolution applied to biology, is "that the innumerable species, genera, and families of organic beings with which the world is peopled have all descended, each within its own class or group, from common parents, and have all been modified in the course of descent."[1]

And, in view of the facts of geology, it follows that all living animals and plants "are the lineal descendants of those which lived long before the Silurian epoch."[2]

It is an obvious consequence of this theory of descent with modification, as it is sometimes called, that all plants and animals, however different they may now be, must, at one time or other, have been connected by direct or indirect intermediate gradations, and that the appearance of isolation presented by various groups of organic beings must be unreal.

No part of Mr. Darwin's work ran more directly counter to the prepossessions of naturalists twenty years ago than this. And such prepossessions were very excusable, for there was undoubtedly a great deal to be said, at that time, in favour of the fixity of

[1] "Origin of Species," ed. 1, p. 457.
[2] *Ibid.* p. 458.

species and of the existence of great breaks, which there was no obvious or probable means of filling up, between various groups of organic beings.

For various reasons, scientific and unscientific, much had been made of the hiatus between man and the rest of the higher mammalia, and it is no wonder that issue was first joined on this part of the controversy. I have no wish to revive past and happily forgotten controversies; but I must state the simple fact that the distinctions in the cerebral and other characters, which were so hotly affirmed to separate man from all other animals in 1860, have all been demonstrated to be non-existent, and that the contrary doctrine is now universally accepted and taught.

But there were other cases in which the wide structural gaps asserted to exist between one group of animals and another were by no means fictitious; and, when such structural breaks were real, Mr. Darwin could account for them only by supposing that the intermediate forms which once existed had become extinct. In a remarkable passage he says—

"We may thus account even for the distinctness of whole classes from each other—for instance, of birds from all other vertebrate animals—by the belief that many animal forms of life have been utterly lost, through which the early progenitors of birds were formerly connected with the early progenitors of the other vertebrate classes."[1]

Adverse criticism made merry over such suggestions as these. Of course it was easy to get out of

[1] "Origin of the Species," p. 431.

the difficulty by supposing extinction; but where was the slightest evidence that such intermediate forms between birds and reptiles as the hypothesis required ever existed? And then probably followed a tirade upon this terrible forsaking of the paths of "Baconian induction."

But the progress of knowledge has justified Mr. Darwin to an extent which could hardly have been anticipated. In 1862, the specimen of *Archæopteryx*, which, until the last two or three years, has remained unique, was discovered; and it is an animal which, in its feathers and the greater part of its organisation, is a veritable bird, while, in other parts, it is as distinctly reptilian.

In 1868, I had the honour of bringing under your notice, in this theatre, the results of investigations made, up to that time, into the anatomical characters of certain ancient reptiles, which showed the nature of the modifications in virtue of which the type of the quadrupedal reptile passed into that of a bipedal bird; and abundant confirmatory evidence of the justice of the conclusions which I then laid before you has since come to light.

In 1875, the discovery of the toothed birds of the cretaceous formation in North America by Professor Marsh completed the series of transitional forms between birds and reptiles, and removed Mr. Darwin's proposition that "many animal forms of life have been utterly lost, through which the early progenitors of birds were formerly connected with the early progenitors of the other vertebrate classes," from the region of hypothesis to that of demonstrable fact.

In 1859, there appeared to be a very sharp and clear hiatus between vertebrated and invertebrated animals, not only in their structure, but, what was more important, in their development. I do not think that we even yet know the precise links of connection between the two; but the investigations of Kowalewsky and others upon the development of *Amphioxus* and of the *Tunicata* prove, beyond a doubt, that the differences which were supposed to constitute a barrier between the two are non-existent. There is no longer any difficulty in understanding how the vertebrate type may have arisen from the invertebrate, though the full proof of the manner in which the transition was actually effected may still be lacking.

Again, in 1859, there appeared to be a no less sharp separation between the two great groups of flowering and flowerless plants. It is only subsequently that the series of remarkable investigations inaugurated by Hofmeister has brought to light the extraordinary and altogether unexpected modifications of the reproductive apparatus in the *Lycopodiaceæ*, the *Rhizocarpeæ*, and the *Gymnospermeæ*, by which the ferns and the mosses are gradually connected with the Phanerogamic division of the vegetable world.

So, again, it is only since 1859 that we have acquired that wealth of knowledge of the lowest forms of life which demonstrates the futility of any attempt to separate the lowest plants from the lowest animals, and shows that the two kingdoms of living nature have a common borderland which belongs to both or to neither.

Thus it will be observed that the whole tendency of biological investigation, since 1859, has been in the direction of removing the difficulties which the apparent breaks in the series created at that time; and the recognition of gradation is the first step towards the acceptance of evolution.

As another great factor in bringing about the change of opinion which has taken place among naturalists, I count the astonishing progress which has been made in the study of embryology. Twenty years ago, not only were we devoid of any accurate knowledge of the mode of development of many groups of animals and plants, but the methods of investigation were rude and imperfect. At the present time, there is no important group of organic beings the development of which has not been carefully studied; and the modern methods of hardening and section-making enable the embryologist to determine the nature of the process, in each case, with a degree of minuteness and accuracy which is truly astonishing to those whose memories carry them back to the beginnings of modern histology. And the results of these embryological investigations are in complete harmony with the requirements of the doctrine of evolution. The first beginnings of all the higher forms of animal life are similar, and however diverse their adult conditions, they start from a common foundation. Moreover, the process of development of the animal or the plant from its primary egg or germ is a true process of evolution—a progress from almost formless to more or less highly organised matter, in virtue of the properties inherent in that matter.

To those who are familiar with the process of development, all *à priori* objections to the doctrine of biological evolution appear childish. Any one who has watched the gradual formation of a complicated animal from the protoplasmic mass, which constitutes the essential element of a frog's or a hen's egg, has had under his eyes sufficient evidence that a similar evolution of the whole animal world from the like foundation is, at any rate, possible.

Yet another product of investigation has largely contributed to the removal of the objections to the doctrine of evolution current in 1859. It is the proof afforded by successive discoveries that Mr. Darwin did not over-estimate the imperfection of the geological record. No more striking illustration of this is needed than a comparison of our knowledge of the mammalian fauna of the Tertiary epoch in 1859 with its present condition. M. Gaudry's researches on the fossils of Pikermi were published in 1868, those of Messrs. Leidy, Marsh, and Cope, on the fossils of the Western Territories of America, have appeared almost wholly since 1870, those of M. Filhol on the phosphorites of Quercy in 1878. The general effect of these investigations has been to introduce to us a multitude of extinct animals, the existence of which was previously hardly suspected; just as if zoologists were to become acquainted with a country, hitherto unknown, as rich in novel forms of life as Brazil or South Africa once were to Europeans. Indeed, the fossil fauna of the Western Territories of America bids fair to exceed in interest and importance all other known Tertiary deposits put together; and yet, with the exception of the case of the American tertiaries, these investigations have extended over very limited areas; and, at Pikermi, were confined to an extremely small space.

Such appear to me to be the chief events in the history of the progress of knowledge during the last twenty years, which account for the changed feeling with which the doctrine of evolution is at present regarded by those who have followed the advance of biological science, in respect of those problems which bear indirectly upon that doctrine.

But all this remains mere secondary evidence. It may remove dissent, but it does not compel assent. Primary and direct evidence in favour of evolution can be furnished only by palæontology. The geological record, so soon as it approaches completeness, must, when properly questioned, yield either an affirmative or a negative answer: if evolution has taken place, there will its mark be left; if it has not taken place, there will lie its refutation.

What was the state of matters in 1859? Let us hear Mr. Darwin, who may be trusted always to state the case against himself as strongly as possible.

"On this doctrine of the extermination of an

infinitude of connecting links between the living and extinct inhabitants of the world, and at each successive period between the extinct and still older species, why is not every geological formation charged with such links? Why does not every collection of fossil remains afford plain evidence of the gradation and mutation of the forms of life? We meet with no such evidence, and this is the most obvious and plausible of the many objections which may be urged against my theory."[1]

Nothing could have been more useful to the opposition than this characteristically candid avowal, twisted as it immediately was into an admission that the writer's views were contradicted by the facts of palæontology. But, in fact, Mr. Darwin made no such admission. What he says in effect is, not that palæontological evidence is against him, but that it is not distinctly in his favour; and, without attempting to attenuate the fact, he accounts for it by the scantiness and the imperfection of that evidence.

What is the state of the case now, when, as we have seen, the amount of our knowledge respecting the mammalia of the Tertiary epoch is increased fifty-fold, and in some directions even approaches completeness?

Simply this, that, if the doctrine of evolution had not existed, palæontologists must have invented it, so irresistibly is it forced upon the mind by the study of the remains of the Tertiary mammalia which have been brought to light since 1859.

[1] "Origin of Species," ed. 1, p. 463.

Among the fossils of Pikermi, Gaudry found the successive stages by which the ancient civets passed into the more modern hyænas; through the Tertiary deposits of Western America, Marsh tracked the successive forms by which the ancient stock of the horse has passed into its present form; and innumerable less complete indications of the mode of evolution of other groups of the higher mammalia have been obtained. In the remarkable memoir on the phosphorites of Quercy, to which I have referred, M. Filhol describes no fewer than seventeen varieties of the genus *Cynodictis*, which fill up all the interval between the viverine animals and the bear-like dog *Amphicyon*; nor do I know any solid ground of objection to the supposition that, in this *Cynodictis-Amphicyon* group, we have the stock whence all the Viveridæ, Felidæ, Hyænidæ, Canidæ, and perhaps the Procyonidæ and Ursidæ, of the present fauna have been evolved. On the contrary, there is a great deal to be said in favour.

In the course of summing up his results, M. Filhol observes:—

"During the epoch of the phosphorites, great changes took place in animal forms, and almost the same types as those which now exist became defined from one another.

"Under the influence of natural conditions of which we have no exact knowledge, though traces of them are discoverable, species have been modified in a thousand ways: races have arisen which, becoming fixed, have thus produced a corresponding number of secondary species."

In 1859, language of which this is an unintentional paraphrase, occurring in the "Origin of Species," was scouted as wild speculation; at present, it is a sober statement of the conclusions to which an acute and critically-minded investigator is led by large and patient study of the facts of palæontology. I venture to repeat what I have said before, that, so far as the animal world is concerned, evolution is no longer a speculation, but a statement of historical fact. It takes its place alongside of those accepted truths which must be reckoned with by philosophers of all schools.

Thus when, on the first day of October next, the "Origin of Species" comes of age, the promise of its youth will be amply fulfilled; and we shall be prepared to congratulate the venerated author of the book, not only that the greatness of his achievement and its enduring influence upon the progress of knowledge have won him a place beside our Harvey; but, still more, that, like Harvey, he has lived long enough to outlast detraction and opposition, and to see the stone that the builders rejected become the head-stone of the corner.

XIII.

THE CONNECTION OF THE BIOLOGICAL SCIENCES WITH MEDICINE.

THE great body of theoretical and practical knowledge which has been accumulated by the labours of some eighty generations, since the dawn of scientific thought in Europe, has no collective English name to which an objection may not be raised; and I use the term "medicine" as that which is least likely to be misunderstood; though, as every one knows, the name is commonly applied, in a narrower sense, to one of the chief divisions of the totality of medical science.

Taken in this broad sense, "medicine" not merely denotes a kind of knowledge, but it comprehends the various applications of that knowledge to the alleviation of the sufferings, the repair of the injuries, and the conservation of the health, of living beings. In fact, the practical aspect of medicine so far dominates over every other, that the "Healing Art" is one of its most widely-received synonyms. It is so difficult to think of medicine otherwise than as something which is necessarily connected with curative treatment, that we are apt to forget that there must be, and is, such a thing as a pure science of medicine—a "pathology" which has no more necessary subservience to practical ends than has zoology or botany.

The logical connection between this purely scientific doctrine of disease, or pathology, and ordinary biology, is easily traced. Living matter is characterised by its innate tendency to exhibit a definite series of the morphological and physiological phenomena which constitute organisation and life. Given a certain range of conditions, and these phenomena remain the same, within narrow limits, for each kind of living thing. They furnish the normal and typical character of the species, and, as such, they are the subject-matter of ordinary biology.

Outside the range of these conditions, the normal course of the cycle of vital phenomena is disturbed; abnormal structure makes its appearance, or the proper character and mutual adjustment of the functions cease to be preserved. The extent and the importance of these deviations from the typical life may vary indefinitely. They may have no noticeable influence on the general well-being of the economy, or they may favour it. On the other hand, they may be of such a nature as to impede the activities of the organism, or even to involve its destruction.

In the first case, these perturbations are ranged under the wide and somewhat vague category of "variations;" in the second, they are called lesions, states of poisoning, or diseases; and, as morbid states, they lie within the province of pathology. No sharp line of demarcation can be drawn between the two classes of phenomena. No one can say where anatomical variations end and tumours begin, nor where modification of function, which may at first promote health, passes into disease. All that can be said is, that whatever change of structure or function is hurtful belongs to pathology. Hence it is obvious that pathology is a branch of biology; it is the morphology, the physiology, the distribution, the ætiology of abnormal life.

However obvious this conclusion may be now, it was nowise apparent in the infancy of medicine. For it is a peculiarity of the physical sciences, that they are independent in proportion as they are imperfect; and it is only as they advance that the bonds which really unite them all become apparent. Astronomy had no manifest connection with terrestrial physics before the publication of the "Principia;" that of chemistry with physics is of still more modern revelation; that of physics and chemistry with physiology, has been stoutly denied within the recollection of most of us, and perhaps still may be.

Or, to take a case which affords a closer parallel with that of medicine. Agriculture has been cultivated from the earliest times, and, from a remote antiquity, men have attained considerable practical skill in the cultivation of the useful plants, and have empirically established many scientific truths concerning the conditions under which they flourish. But, it is within the memory of many of us, that chemistry on the one hand, and vegetable physiology on the other, attained a stage of development such that they were able to furnish a sound basis for scientific agriculture. Similarly, medicine took its rise in the practical needs of mankind. At first, studied without reference to any other branch of knowledge, it long maintained, indeed still to some extent maintains, that independence. Historically, its connection with the biological sciences has been slowly established, and the full extent and intimacy of that connection are only now beginning to be apparent. I trust I have not been mistaken in supposing that an attempt to give a brief sketch of the steps by which a philosophical necessity has become an historical reality, may not be devoid of interest, possibly of instruction, to the members of this great Congress, profoundly interested as all are in the scientific development of medicine.

The history of medicine is more complete and fuller than that of any other science, except, perhaps, astronomy; and, if we follow back the long record as far as clear evidence lights us, we find ourselves taken to the early stages of the civilisation of Greece. The oldest hospitals were the temples of Æsculapius; to these Asclepeia, always erected on healthy sites, hard by fresh springs and surrounded by shady groves, the sick and the maimed resorted to seek the aid of the god of health. Votive tablets or inscriptions recorded the symptoms, no less than the gratitude, of those who were healed; and, from these primitive clinical records, the half-priestly, half-philosophic caste of the Asclepiads compiled the data upon which the earliest generalisations of medicine, as an inductive science, were based.

In this state, pathology, like all the inductive sciences at their origin, was merely natural history; it registered the phenomena of disease, classified them, and ventured upon a prognosis, wherever the observation of constant co-existences and sequences suggested a rational expectation of the like recurrence under similar circumstances.

Further than this it hardly went. In fact, in the then state of knowledge, and in the condition of philosophical speculation at that time, neither the causes of the morbid state, nor the *rationale* of treatment, were likely to be sought for as we seek for them now. The anger of a god was a sufficient reason for the existence of a malady, and a dream ample warranty for therapeutic measures; that a physical phenomenon must needs have a physical cause was not the implied or expressed axiom that it is to us moderns.

The great man whose name is inseparately connected with the foundation of medicine, Hippocrates, certainly knew very little, indeed practically nothing, of anatomy or physiology; and he would, probably, have been perplexed, even to imagine the possibility of a connection between the zoological studies of his contemporary Democritus and medicine. Nevertheless, in so far as he, and those who worked before and after him, in the same spirit, ascertained, as matters of experience, that a wound, or a luxation, or a fever, presented such and such symptoms, and that the return of the patient to health was facilitated by such and such measures, they established laws of nature, and began the construction of the science of pathology. All true science begins with empiricism—though all true science is such exactly, in so far as it strives to pass out of the empirical stage into that of the deduction of empirical from more general truths. Thus, it is not wonderful, that the early physicians had little or nothing to do with the development of biological science; and, on the other hand, that the early biologists did not much concern themselves with medicine. There is nothing to show that the Asclepiads took any prominent share in the work of founding anatomy, physiology, zoology, and botany. Rather do these seem to have sprung from the early philosophers, who were essentially natural philosophers, animated by the characteristically Greek thirst for knowledge as such. Pythagoras, Alcmeon, Democritus, Diogenes of Apollonia, are all credited with anatomical and physiological investigations; and, though Aristotle is said to have belonged to an Asclepiad family, and not improbably owed his taste for anatomical and zoological inquiries to the teachings of his father, the physician Nicomachus, the "Historia Animalium," and the treatise "De Partibus Animalium," are as free from any allusion to medicine as if they had issued from a modern biological laboratory.

It may be added, that it is not easy to see in what way it could have benefited a physician of Alexander's time to know all that Aristotle knew on these subjects. His human anatomy was too rough to avail much in diagnosis; his physiology was too erroneous to supply data for pathological reasoning. But when the Alexandrian school, with Erasistratus and Herophilus at their head, turned to account the opportunities of studying human structure, afforded to them by the Ptolemies, the value of the large amount of accurate knowledge thus obtained to the surgeon for his operations, and to the physician for his diagnosis of internal disorders, became obvious, and a connection was established between anatomy and medicine, which has ever become closer and closer. Since the revival of learning, surgery, medical diagnosis, and anatomy have gone hand in hand. Morgagni called his great work, "De sedibus et causis morborum per anatomen indagatis," and not only showed the way to search out the localities and the causes of disease by anatomy, but himself travelled wonderfully far upon the road. Bichat, discriminating the grosser constituents of the organs and parts of the body, one from another, pointed out the direction which modern research must take; until, at length, histology, a science of yesterday, as it seems to many of us, has carried the work of Morgagni as far as the microscope can take us, and has extended the realm of pathological anatomy to the limits of the invisible world.

Thanks to the intimate alliance of morphology with medicine, the natural history of disease has, at the present day, attained a high degree of perfection. Accurate regional anatomy has rendered practicable the exploration of the most hidden parts of the organism, and the determination, during life, of morbid changes in them; anatomical and histological postmortem investigations have supplied physicians with a clear basis upon which to rest the classification of diseases, and with unerring tests of the accuracy or inaccuracy of their diagnoses.

If men could be satisfied with pure knowledge, the extreme precision with which, in these days, a sufferer may be told what is happening, and what is likely to happen, even in the most recondite parts of his bodily frame, should be as satisfactory to the patient as it is to the scientific pathologist who gives him the information. But I am afraid it is not; and even the practising physician, while nowise underestimating the regulative value of accurate diagnosis, must often lament that so much of his knowledge rather prevents him from doing wrong than helps him to do right.

A scorner of physic once said that nature and disease may be compared to two men fighting, the doctor to a blind man with a club, who strikes into the *melée*, sometimes hitting the disease, and sometimes hitting nature. The matter is not mended if you suppose the blind man's hearing to be so acute that he can register every stage of the struggle, and pretty clearly predict how it will end. He had better not meddle at all, until his eyes are opened—until he can see the exact position of the antagonists, and make sure of the effect of his blows. But that which it behoves the physician to see, not, indeed, with his bodily eye, but with clear, intellectual vision, is a process, and the chain of causation involved in that process. Disease, as we have seen, is a perturbation of the normal activities of a living body, and it is, and must remain, unintelligible, so long as we are ignorant of the nature of these normal activities. In other words, there could be no real science of pathology until the science of physiology had reached a degree of perfection unattained, and indeed unattainable, until quite recent times.

So far as medicine is concerned, I am not sure that physiology, such as it was down to the time of Harvey, might as well not have existed. Nay, it is perhaps no exaggeration to say that, within the memory of living men, justly renowned practitioners of medicine and surgery knew less physiology than is now to be learned from the most elementary text-book; and, beyond a few broad facts, regarded what they did know as of extremely little practical importance. Nor am I disposed to blame them for this conclusion; physiology must be useless, or worse than useless, to pathology, so long as its fundamental conceptions are erroneous.

Harvey is often said to be the founder of modern physiology; and there can be no question that the elucidations of the function of the heart, of the nature of the pulse, and of the course of the blood, put forth in the ever-memorable little essay, "De motu cordis," directly worked a revolution in men's views of the nature and of the concatenation of some of the most important physiological processes among the higher animals; while, indirectly, their influence was perhaps even more remarkable.

But, though Harvey made this signal and perennially important contribution to the physiology of the moderns, his general conception of vital processes was essentially identical with that of the ancients; and, in the "Exercitationes de generatione," and notably in the singular chapter "De calido innato," he shows himself a true son of Galen and of Aristotle.

For Harvey, the blood possesses powers superior to those of the elements; it is the seat of a soul which is not only vegetative, but also sensitive and motor. The blood maintains and fashions all parts of the body, "idque summâ cum providentiâ et intellectu in finem certum agens, quasi ratiocinio quodam uteretur."

Here is the doctrine of the "pneuma," the product of the philosophical mould into which the animism of primitive men ran in Greece, in full force. Nor did its strength abate for long after Harvey's time. The same ingrained tendency of the human mind to suppose that a process is explained when it is ascribed to a power of which nothing is known except that it is the hypothetical agent of the process, gave rise, in the next century, to the animism of Stahl; and, later, to the doctrine of a vital principle, that "asylum ignorantiæ" of physiologists, which has so easily accounted for everything and explained nothing, down to our own times.

Now the essence of modern, as contrasted with ancient, physiological science appears to me to lie in its antagonism to animistic hypotheses and animistic phraseology. It offers physical explanations of vital phenomena, or frankly confesses that it has none to offer. And, so far as I know, the first person who gave expression to this modern view of physiology, who was bold enough to enunciate the proposition that vital phenomena, like all the other phenomena of the physical world, are, in ultimate analysis, resolvable into matter and motion, was René Descartes.

The fifty-four years of life of this most original and powerful thinker are widely overlapped, on both sides, by the eighty of Harvey, who survived his younger contemporary by seven years, and takes pleasure in acknowledging the French philosopher's appreciation of his great discovery.

In fact, Descartes accepted the doctrine of the circulation as propounded by "Harvæus médecin d'Angleterre," and gave a full account of it in his first work, the famous "Discours de la Méthode," which was published in 1637, only nine years after the exercitation "De motu cordis;" and, though differing from Harvey on some important points (in which it may be noted, in passing, Descartes was wrong and Harvey right), he always speaks of him with great respect. And so important does the subject seem to Descartes, that he returns to it in the "Traité des Passions," and in the "Traité de l'Homme."

It is easy to see that Harvey's work must have had a peculiar significance for the subtle thinker, to whom we owe both the spiritualistic and the materialistic philosophies of modern times. It was in the very year of its publication, 1628, that Descartes withdrew into that life of solitary investigation and meditation of which his philosophy was the fruit. And, as the course of his speculations led him to establish an absolute distinction of nature between the material and the mental worlds, he was logically compelled to seek for the explanation of the phenomena of the material world within itself; and having allotted the realm of thought to the soul, to see nothing but extension and motion in the rest of nature. Descartes uses "thought" as the equivalent of our modern term "consciousness." Thought is the function of the soul, and its only function. Our natural heat and all the movements of the body, says he, do not depend on the soul. Death does not take place from any fault of the soul, but only because some of the principal parts of the body become corrupted. The body of a living man differs from that of a dead man in the same way as a watch or other automaton (that is to say, a machine which moves of itself) when it is wound up and has, in itself, the physical principle of the movements which the mechanism is adapted to perform, differs from the same watch, or other machine, when it is broken, and the physical principle of its movement no longer exists. All the actions which are common to us and the lower animals depend only on the conformation of our organs, and the course which the animal spirits take in the brain, the nerves, and the muscles; in the same way as the movement of a watch is produced by nothing but the force of its spring and the figure of its wheels and other parts.

Descartes' "Treatise on Man" is a sketch of human physiology, in which a bold attempt is made to explain all the phenomena of life, except those of consciousness, by physical reasonings. To a mind turned in this direction, Harvey's exposition of the heart and vessels as a hydraulic mechanism must have been supremely welcome.

Descartes was not a mere philosophical theorist, but a hardworking dissector and experimenter, and he held the strongest opinion respecting the practical value of the new conception which he was introducing. He speaks of the importance of preserving health, and of the dependence of the mind on the body being so close that, perhaps, the only way of making men wiser and better than they are, is to be sought in medical science. "It is true," says he, "that as medicine is now practised, it contains little that is very useful; but without any desire to depreciate, I am sure that there is no one, even among professional men, who will not declare that all we know is very little as compared with that which remains to be known; and that we might escape an infinity of diseases of the mind, no less than of the body, and even perhaps from the weakness of old age, if we had sufficient knowledge of their causes, and of all the remedies with which nature has provided us."[1] So strongly impressed was Descartes with this, that he resolved to spend the rest of his life in trying to acquire such a knowledge of nature as would lead to the construction of a better medical doctrine.[2] The anti-Cartesians found material for cheap ridicule in these aspirations of the philosopher; and it is almost needless to say that, in the thirteen years which elapsed between the publication of the "Discours" and

1 "Discours de la Méthode," 6e partie, Ed. Cousin, p. 193

2 *Ibid.* pp. 193 and 211.

the death of Descartes, he did not contribute much to their realisation. But, for the next century, all progress in physiology took place along the lines which Descartes laid down.

The greatest physiological and pathological work of the seventeenth century, Borelli's treatise "De Motu Animalium," is, to all intents and purposes, a development of Descartes' fundamental conception; and the same may be said of the physiology and pathology of Boerhaave, whose authority dominated in the medical world of the first half of the eighteenth century.

With the origin of modern chemistry, and of electrical science, in the latter half of the eighteenth century, aids in the analysis of the phenomena of life, of which Descartes could not have dreamed, were offered to the physiologist. And the greater part of the gigantic progress which has been made in the present century is a justification of the prevision of Descartes. For it consists, essentially, in a more and more complete resolution of the grosser organs of the living body into physico-chemical mechanisms.

"I shall try to explain our whole bodily machinery in such a way, that it will be no more necessary for us to suppose that the soul produces such movements as are not voluntary, than it is to think that there is in a clock a soul which causes it to show the hours."[1] These words of Descartes might be appropriately taken as a motto by the author of any modern treatise on physiology.

[1] "De la Formation du Fœtus."

But though, as I think, there is no doubt that Descartes was the first to propound the fundamental conception of the living body as a physical mechanism, which is the distinctive feature of modern, as contrasted with ancient physiology, he was misled by the natural temptation to carry out, in all its details, a parallel between the machines with which he was familiar, such as clocks and pieces of hydraulic apparatus, and the living machine. In all such machines there is a central source of power, and the parts of the machine are merely passive distributors of that power. The Cartesian school conceived of the living body as a machine of this kind; and herein they might have learned from Galen, who, whatever ill use he may have made of the doctrine of "natural faculties," nevertheless had the great merit of perceiving that local forces play a great part in physiology.

The same truth was recognised by Glisson, but it was first prominently brought forward in the Hallerian doctrine of the "vis insita" of muscles. If muscle can contract without nerve, there is an end of the Cartesian mechanical explanation of its contraction by the influx of animal spirits.

The discoveries of Trembley tended in the same direction. In the freshwater *Hydra,* no trace was to be found of that complicated machinery upon which the performance of the functions in the higher animals was supposed to depend. And yet the hydra moved, fed, grew, multiplied, and its fragments exhibited all the powers of the whole. And, finally, the work of Caspar F. Wolff,[1] by demonstrating the fact that the growth and development of both plants and animals take place antecedently to the existence of their grosser organs, and are, in fact, the causes and not the consequences of organisation (as then understood), sapped the foundations of the Cartesian physiology as a complete expression of vital phenomena.

For Wolff, the physical basis of life is a fluid, possessed of a "vis essentialis" and a "solidescibilitas," in virtue of which it gives rise to organisation; and, as he points out, this conclusion strikes at the root of the whole iatro-mechanical system.

In this country, the great authority of John Hunter exerted a similar influence; though it must be admitted that the too sibylline utterances which are the outcome of Hunter's struggles to define his conceptions are often susceptible of more than one interpretation. Nevertheless, on some points Hunter is clear enough. For example, he is of opinion that "Spirit is only a property of matter" ("Introduction to Natural History," p. 6), he is prepared to renounce animism (*l.c.* p. 8), and his conception of life is so completely physical that he thinks of it as something which can exist in a state of combination in the food. "The aliment we take in has in it, in a fixed state, the real life; and this does not become active until it has got into the lungs; for there it is freed from its prison" ("Observations on Physiology," p. 113). He also thinks that "It is more in accord with the

1 "Theoria Generationis," 1759.

general principles of the animal machine to suppose that none of its effects are produced from any mechanical principle whatever; and that every effect is produced from an action in the part; which action is produced by a stimulus upon the part which acts, or upon some other part with which this part sympathises so as to take up the whole action" (*l.c.* p. 152).

And Hunter is as clear as Wolff, with whose work he was probably unacquainted, that "whatever life is, it most certainly does not depend upon structure or organisation" (*l.c.* p. 114).

Of course it is impossible that Hunter could have intended to deny the existence of purely mechanical operations in the animal body. But while, with Borelli and Boerhaave, he looked upon absorption, nutrition, and secretion as operations effected by means of the small vessels, he differed from the mechanical physiologists, who regarded these operations as the result of the mechanical properties of the small vessels, such as the size, form, and disposition of their canals and apertures. Hunter, on the contrary, considers them to be the effect of properties of these vessels which are not mechanical but vital. "The vessels," says he, "have more of the polypus in them than any other part of the body," and he talks of the "living and sensitive principles of the arteries," and even of the "dispositions or feelings of the arteries." "When the blood is good and genuine the sensations of the arteries, or the dispositions for sensation, are agreeable. . . . It is then they dispose of the blood to the best advantage, increasing the growth of the whole, supplying any losses, keeping up a due succession, etc." (*l.c.* p. 133).

If we follow Hunter's conceptions to their logical issue, the life of one of the higher animals is essentially the sum of the lives of all the vessels, each of which is a sort of physiological unit, answering to a polype; and, as health is the result of the normal "action of the vessels," so is disease an effect of their abnormal action. Hunter thus stands in thought, as in time, midway between Borelli on the one hand, and Bichat on the other.

The acute founder of general anatomy, in fact, outdoes Hunter in his desire to exclude physical reasonings from the realm of life. Except in the interpretation of the action of the sense organs, he will not allow physics to have anything to do with physiology.

"To apply the physical sciences to physiology is to explain the phenomena of living bodies by the laws of inert bodies. Now this is a false principle, hence all its consequences are marked with the same stamp. Let us leave to chemistry its affinity; to physics, its elasticity and its gravity. Let us invoke for physiology only sensibility and contractility."[1]

Of all the unfortunate dicta of men of eminent ability this seems one of the most unhappy, when we think of what the application of the methods and the data of physics and chemistry has done towards bringing physiology into its present state. It is not

1 "Anatomie générale," i. p. liv.

too much to say that one half of a modern text-book of physiology consists of applied physics and chemistry; and that it is exactly in the exploration of the phenomena of sensibility and contractility that physics and chemistry have exerted the most potent influence.

Nevertheless, Bichat rendered a solid service to physiological progress by insisting upon the fact that what we call life, in one of the higher animals, is not an indivisible unitary archæus dominating, from its central seat, the parts of the organism, but a compound result of the synthesis of the separate lives of those parts.

"All animals," says he, "are assemblages of different organs, each of which performs its function and concurs, after its fashion, in the preservation of the whole. They are so many special machines in the general machine which constitutes the individual. But each of these special machines is itself compounded of many tissues of very different natures, which in truth constitute the elements of those organs" (*l.c.* lxxix.) "The conception of a proper vitality is applicable only to these simple tissues, and not to the organs themselves" (*l.c.* lxxxiv.)

And Bichat proceeds to make the obvious application of this doctrine of synthetic life, if I may so call it, to pathology. Since diseases are only alterations of vital properties, and the properties of each tissue are distinct from those of the rest, it is evident that the diseases of each tissue must be different from those of the rest. Therefore, in any organ composed of different tissues, one may be diseased and the other remain healthy; and this is what happens in most cases (*l.c.* lxxxv.)

In a spirit of true prophecy, Bichat says, "We have arrived at an epoch, in which pathological anatomy should start afresh." For, as the analysis of the organs had led him to the tissues, as the physiological units of the organism; so, in a succeeding generation, the analysis of the tissues led to the cell as the physiological element of the tissues. The contemporaneous study of development brought out the same result; and the zoologists and botanists, exploring the simplest and the lowest forms of animated beings, confirmed the great induction of the cell theory. Thus the apparently opposed views, which have been battling with one another ever since the middle of the last century, have proved to be each half the truth.

The proposition of Descartes that the body of a living man is a machine, the actions of which are explicable by the known laws of matter and motion, is unquestionably largely true. But it is also true, that the living body is a synthesis of innumerable physiological elements, each of which may nearly be described, in Wolff's words, as a fluid possessed of a "vis essentialis," and a "solidescibilitas"; or, in modern phrase, as protoplasm susceptible of structural metamorphosis and functional metabolism: and that the only machinery, in the precise sense in which the Cartesian school understood mechanism, is, that which co-ordinates and regulates these physiological units into an organic whole.

In fact, the body is a machine of the nature of an

army, not of that of a watch or of a hydraulic apparatus. Of this army each cell is a soldier, an organ a brigade, the central nervous system headquarters and field telegraph, the alimentary and circulatory system the commissariat. Losses are made good by recruits born in camp, and the life of the individual is a campaign, conducted successfully for a number of years, but with certain defeat in the long run.

The efficacy of an army, at any given moment, depends on the health of the individual soldier, and on the perfection of the machinery by which he is led and brought into action at the proper time; and, therefore, if the analogy holds good, there can be only two kinds of diseases, the one dependent on abnormal states of the physiological units, the other on perturbations of their co-ordinating and alimentative machinery.

Hence, the establishment of the cell theory, in normal biology, was swiftly followed by a "cellular pathology," as its logical counterpart. I need not remind you how great an instrument of investigation this doctrine has proved in the hands of the man of genius to whom its development is due, and who would probably be the last to forget that abnormal conditions of the co-ordinative and distributive machinery of the body are no less important factors of disease.

Henceforward, as it appears to me, the connection of medicine with the biological sciences is clearly defined. Pure pathology is that branch of biology which defines the particular perturbation of cell-life, or of the co-ordinating machinery, or of both, on which the phenomena of disease depend.

Those who are conversant with the present state of biology will hardly hesitate to admit that the conception of the life of one of the higher animals as the summation of the lives of a cell aggregate, brought into harmonious action by a co-ordinative machinery formed by some of these cells, constitutes a permanent acquisition of physiological science. But the last form of the battle between the animistic and the physical views of life is seen in the contention whether the physical analysis of vital phenomena can be carried beyond this point or not.

There are some to whom living protoplasm is a substance, even such as Harvey conceived the blood to be, "summâ cum providentiâ et intellectu in finem certum agens, quasi ratiocinio quodam;" and who look with as little favour as Bichat did, upon any attempt to apply the principles and the methods of physics and chemistry to the investigation of the vital processes of growth, metabolism, and contractility. They stand upon the ancient ways; only, in accordance with that progress towards democracy, which a great political writer has declared to be the fatal characteristic of modern times, they substitute a republic formed by a few billion of "animulæ" for the monarchy of the all-pervading "anima."

Others, on the contrary, supported by a robust faith in the universal applicability of the principles laid down by Descartes, and seeing that the actions called "vital" are, so far as we have any means of knowing, nothing but changes of place of particles of matter, look to molecular physics to achieve the analysis of the living protoplasm itself into a molecular mechanism. If there is any truth in the received doctrines of physics, that contrast between living and inert matter, on which Bichat lays so much stress, does not exist. In nature, nothing is at rest, nothing is amorphous; the simplest particle of that which men in their blindness are pleased to call "brute matter" is a vast aggregate of molecular mechanisms performing complicated movements of immense rapidity, and sensitively adjusting themselves to every change in the surrounding world. Living matter differs from other matter in degree and not in kind; the microcosm repeats the macrocosm; and one chain of causation connects the nebulous original of suns and planetary systems with the protoplasmic foundation of life and organisation.

From this point of view, pathology is the analogue of the theory of perturbations in astronomy; and therapeutics resolves itself into the discovery of the means by which a system of forces competent to eliminate any given perturbation may be introduced into the economy. And, as pathology bases itself upon normal physiology, so therapeutics rests upon pharmacology; which is, strictly speaking, a part of the great biological topic of the influence of conditions on the living organism, and has no scientific foundation apart from physiology.

It appears to me that there is no more hopeful indication of the progress of medicine towards the ideal of Descartes than is to be derived from a comparison of the state of pharmacology, at the present day, with that which existed forty years ago. If we consider the knowledge positively acquired, in this short time, of the *modus operandi* of urari, of atropia, of physostigmin, of veratria, of casca, of strychnia, of bromide of potassium, of phosphorus, there can surely be no ground for doubting that, sooner or later, the pharmacologist will supply the physician with the means of affecting, in any desired sense, the functions of any physiological element of the body. It will, in short, become possible to introduce into the economy a molecular mechanism which, like a very cunningly-contrived torpedo, shall find its way to some particular group of living elements, and cause an explosion among them, leaving the rest untouched.

The search for the explanation of diseased states in modified cell-life; the discovery of the important part played by parasitic organisms in the ætiology of disease; the elucidation of the action of medicaments by the methods and the data of experimental physiology; appear to me to be the greatest steps which have ever been made towards the establishment of medicine on a scientific basis. I need hardly say they could not have been made except for the advance of normal biology.

There can be no question, then, as to the nature or the value of the connection between medicine and the biological sciences. There can be no doubt that the future of pathology and of therapeutics, and, therefore, that of practical medicine, depends upon the extent to which those who occupy themselves with these subjects are trained in the methods and impregnated with the fundamental truths of biology.

And, in conclusion, I venture to suggest that the collective sagacity of this Congress could occupy itself with no more important question than with this: How is medical education to be arranged, so that, without entangling the student in those details of the systematist which are valueless to him, he may be enabled to obtain a firm grasp of the great truths respecting animal and vegetable life, without which, notwithstanding all the progress of scientific medicine, he will still find himself an empiric?

THE END.

Printed by R. & R. CLARK, *Edinburgh.*

Made in the USA
Monee, IL
04 November 2026

67391568R00203